The Book of

GREEN

Quotations

Edited by
JAMES DALEY

DOVER PUBLICATIONS, INC.
Mineola, New York

Copyright

Copyright © 2009 by Dover Publications, Inc.
All rights reserved.

Bibliographical Note

The Book of Green Quotations is a new work, first published by Dover Publications, Inc., in 2009.

Library of Congress Cataloging-in-Publication Data
The book of green quotations ı edited by James Daley.
   p. cm.
  ISBN-13: 978-0-486-46781-8
  ISBN-10: 0-486-46781-3
   1. Environmentalism—Quotations, maxims, etc. I. Daley, James, 1979–
GE40.B66 2009
333.72—dc22

2008052615

Manufactured in the United States of America
Dover Publications, Inc., 31 East 2nd Street, Mineola, N.Y. 11501

Introduction

IN 1864, GEORGE PERKINS MARSH, perhaps America's first environmentalist, prophetically observed that ". . . Man, who even now finds scarce breathing room on this vast globe, cannot retire from the Old World to some yet undiscovered continent, and wait for the slow action of such causes to replace, by a new creation, the Eden he has wasted." Even then—years before the first automobiles were to drive over the new highways of America, decades before the first power plant would begin burning coal to make electricity, and over a century before the last gallon of oil would spill from the Exxon Valdez into Prince William Sound—it was apparent, at least to some, that mankind's ever-increasing influence over the natural world had the potential to lead to its very destruction.

Yet here we are, in the twenty-first century, still staring down the same barrel that Marsh first faced such a long time ago. The specter of global warming has made the idea of seeing our world become a "wasted Eden" a very real and frightening possibility. Meanwhile, air pollution in our cities and suburbs continues to rise to increasingly toxic levels; the rainforests that provide us with the very air we breathe continue to be depleted; and the natural resources that we depend upon for food, energy, and air come closer to disappearing with every passing day.

Thus, it is with an eye toward this long history of environmental activism that the present volume was compiled. Including quotes that span the centuries, it is the editor's hope that the reader will find within these pages a sense of both the timelessness and the timeliness of the struggle to protect our environment. One can find quotations on the dangers we face now and have faced over the years in the chapters on pollution and global warming; there are quotations about how we fight to overcome these challenges in the sections on environmentalism, ecology, and conservation; and, finally, there are quotations on the beauty and wonder that we are trying to save—and may be lost if we fail—in the sections on nature, the wilderness, and the Earth.

Although some would argue that we have made little progress in the 145 years since Marsh's speech, many more would surely contend that it is not too late to change our present course. As the hundreds of citizens, speakers, world leaders, and activists represented in the following pages can attest, it is nothing short of our moral imperative to create a world that respects our wilderness, protects our environment, and preserves this fragile planet for millennia to come.

JAMES DALEY
Editor

Contents

The Book of
GREEN
Quotations

Conservation

God bless America. Let's save some of it.

EDWARD ABBEY

ℭ

Energy conservation is the foundation of energy independence.

TOM ALLEN

ℭ

Conservation is to a democratic government by free men as the roots of a tree are to its leaves. We must be willing wisely to nurture and use our resources if we are going to keep visible the inner strengths of democracy.

CLINTON P. ANDERSON

ℭ

The conservation conscience can be dominant in American life from time to time and place to place without representing a conviction on the part of the majority of the people.

ANONYMOUS

ℭ

Willful waste brings woeful want.

ANONYMOUS PROVERB

CR

Waste is a tax on the whole people.

ALBERT W. ATWOOD

CR

A true conservationist is a man who knows that the world is not given by his fathers but borrowed from his children.

JOHN JAMES AUDUBON

CR

The American reading his Sunday paper in a state of lazy collapse is perhaps the most perfect symbol of the triumph of quantity over quality. . . . Whole forests are being ground into pulp daily to minister to our triviality.

IRVING BABBITT

CR

If civilization has risen from the Stone Age, it can rise again from the Wastepaper Age.

JACQUES BARZUN

CR

What a country chooses to save is what a country chooses to say about itself.

MOLLIE BEATTIE

CR

We do need a "new economy," but one that is founded on thrift and care, on saving and conserving, not on excess and waste. An economy based on waste is inherently and hopelessly violent, and war is its inevitable by-product. We need a peaceable economy.

We have lived by the assumption that what was good for us would be good for the world. We have been wrong. We must change our lives, so that it will be possible to live by the contrary assumption that what is good for the world will be good for us. . . . We must recover the sense of the majesty of the creation and the ability to be worshipful in its presence. For it is only on the condition of humility and reverence before the world that our species will be able to remain in it.

Our destruction of nature is not just bad stewardship, or stupid economics, or a betrayal of family responsibility; it is the most horrid blasphemy. It is flinging God's gifts into His face, as if they were of no worth beyond that assigned to them by our destruction of them.

WENDELL BERRY

○

The theologians affirm that the conservation of this world is a perpetual creation, and that the verbs "conserve" and "create," so much at odds here, are synonymous in Heaven.

JORGE LUIS BORGES

○

The most unhappy thing about conservation is that it is never permanent. Save a priceless woodland or an irreplaceable mountain today, and tomorrow it is threatened from another quarter.

HAL BORLAND

○

Let man heal the hurt places and revere whatever is still miraculously pristine.

DAVID R. BROWER

○

When a man throws an empty cigarette package from an automobile, he is liable to a fine of $50. When a man throws a billboard across a view, he is richly rewarded.

PAT BROWN

CR

One of the first and most leading principles on which the commonwealth and its laws are consecrated, is lest the temporary possessors and life renters in it, unmindful of what they have received from their ancestors, or of what is due to their posterity, should act as if they were the entire masters; that they should not think it among their rights to cut off the entail, or commit waste on the inheritance, by destroying at their pleasure the whole original fabric of their society; hazarding to those who come after them a ruin instead of a habitation. . . . No one generation could link with another. Men would become little better than flies of a summer.

EDMUND BURKE

CR

Living in the midst of abundance we have the greatest difficulty in seeing that the supply of natural wealth is limited and that the constant increase of population is destined to reduce the American standard of living unless we deal more sanely with our resources.

W. H. CAROTHERS

CR

I want to make it clear if there is ever a conflict (between environmental quality and economic growth), I will go for beauty, clean air, water, and landscape.

JIMMY CARTER

CR

Just as it was impossible to be half a Christian, so too was it impossible to be half a preservationist.

ALSTON CHASE

જ

The air, the water and the ground are free gifts to man, and no one has the power to portion them out in parcels. Man must drink and breathe and walk and therefore each man has a right to his share of each.

JAMES FENIMORE COOPER

જ

We all have an obligation as citizens of this earth to leave the world a healthier, cleaner, and better place for our children and future generations.

BLYTHE DANNER

જ

To the extent that we create or maintain beauty through an ordered diversity, we will also enhance the stability, health, and productivity of America.

RAYMOND F. DASMANN

જ

The use of plant oil as fuel may seem insignificant today; but such products can in time become just as important as kerosene and these coal-tar-products of today.

RUDOLF DIESEL

જ

In the end, we conserve only what we love. We will love only what we understand. We will understand only what we are taught.

BABA DIOUM

ભ

There is a widespread tendency to assume that the
ultimate goal of conservation is to protect nature *against* man,
whereas it should be the discovery and development of the
potentialities that permit a creative harmonious interplay
between man and nature.

RENE DUBOS

ભ

I'd put my money on the sun and solar energy. What a
source of power! I hope we don't have to wait till oil and
coal run out before we tackle that.

THOMAS EDISON

ભ

Since natural resources are finite, increased consumption
must inevitably lead to depletion and scarcity.

PAUL EHRLICH

ભ

The use of sea and air is common to all; neither can a title
to the ocean belong to any people or private persons,
forasmuch as neither nature nor public use and custom
permit any possession thereof.

ELIZABETH I of England

ભ

The sun, the moon and the stars would have disappeared
long ago . . . had they happened to be within the reach of
predatory human hands.

HAVELOCK ELLIS

❧

Conservation is the religion of the future.

JANE FONDA

❧

We are cutting out our kidneys to enlarge our stomachs.

ERIC T. FREYFOGLE

❧

We never know the worth of water till the well is dry.

THOMAS FULLER

❧

Till now man has been up against Nature; from now on he will be up against his own nature.

DENNIS GABOR

❧

God forbid that India should ever take to industrialism after the manner of the west . . . keeping the world in chains. If [our nation] took to similar economic exploitation, it would strip the world bare like locusts.

MOHANDAS GANDHI

❧

The future belongs to those who understand that doing more with less is compassionate, prosperous, and enduring, and thus more intelligent, even competitive.

PAUL HAWKEN

❧

I feel more confident than ever that the power to save the planet rests with the individual consumer.

Listen up, you couch potatoes: each recycled beer can saves enough electricity to run a television for three hours.

DENIS HAYES

෨

We can dream together a dream of a better world, an ever-renewing, organic-based earth-community. I believe we can achieve a large part of it in our lifetime.

RANDY HAYES

෨

At the beginning of the cask and the end take thy fill, but be saving in the middle; for at the bottom the saving comes too late.

HESIOD

෨

The irony of the matter [regarding the environment] is that the future generations do not have a vote. In effect, we hold their proxy.

CHARLES J. HITCH

෨

The people have a vital interest in the conservation of their natural resources; in the prevention of wasteful practices.

I can at once refute the statement that the people of the West object to conservation of oil resources. They know that there is a limit to oil supplies and that the time will come when they and the Nation will need this oil much more than it is needed now. There are no half measures in conservation of oil.

HERBERT HOOVER

⚬

Multiple-purpose development is no longer good enough. All-purpose conservation must be our standard.

HUBERT H. HUMPHREY

⚬

Will urban sprawl spread so far that most people lose all touch with nature? Will the day come when the only bird a typical American child ever sees is a canary in a pet shop window? When the only wild animal he knows is a rat— glimpsed on a night drive through some city slum? When the only tree he touches is the cleverly fabricated plastic evergreen that shades his gifts on Christmas morning?

FRANK N. IKARD

⚬

In our every deliberation, we must consider the impact of our decisions on the next seven generations.

From *The Great Law of the Iroquois Confederacy*

⚬

If future generations are to remember us with gratitude rather than contempt, we must leave them something more than the miracles of technology. We must leave them a glimpse of the world as it was in the beginning, not just after we got through with it.

Conservation is ethically sound. It is rooted in our love of the land, our respect for the rights of others, our devotion to the rule of law.

LYNDON BAINES JOHNSON

☙

Take nothing but pictures.
Leave nothing but footprints.
Kill nothing but time.

JOHN KAY

☙

Nothing is more conservative than conservation.

RUSSELL KIRK

☙

What is commonly called "conservation" will not work in
the long run because it is not really conservation at all but
rather, disguised by its elaborate scheming, only a more
knowledgeable variation of the old idea of a world for man's
use only. That idea is unrealizable.

If people destroy something replaceable made by mankind,
they are called vandals; if they destroy something irreplaceable
made by God, they are called developers.

JOSEPH WOOD KRUTCH

☙

Conservation is a bird that flies faster than the shot we aim
at it.

Conservation is a state of harmony between men and land.

We abuse land because we regard it as a commodity
belonging to us. When we see land as a community to which
we belong, we may begin to use it with love and respect.

To save every cog and wheel is the first precaution of
intelligent tinkering.

Having to squeeze the last drop of utility out of the land has the same desperate finality as having to chop up the furniture to keep warm.

ALDO LEOPOLD

❦

Our ideals, laws and customs should be based on the proposition that each generation, in turn, becomes the custodian rather than the absolute owner of our resources and each generation has the obligation to pass this inheritance on to the future.

CHARLES A. LINDBERGH

❦

Conservation is no longer for the birds.

LES LINE

❦

Conservation is sometimes perceived as stopping every-thing cold, as holding whooping cranes in higher esteem than people. It is up to science to spread the understanding that the choice is not between wild places or people, it is between a rich or an impoverished existence for Man.

THOMAS E. LOVEJOY

❦

I've never been interested in just doing with less. I'm interested in doing *more* with less. We don't have to become vegetarians and ride bicycles to save the Earth.

We'd find more energy in the attics of American homes [through energy conservation measures] than in all the oil buried in Alaska.

AMORY LOVINS

CR

Your grandchildren will likely find it incredible—or even sinful—that you burned up a gallon of gasoline to fetch a pack of cigarettes!

PAUL MACCREADY, JR.

CR

The development of civilization and industry in general has always shown itself so active in the destruction of forests that everything that has been done for their conservation and production is completely insignificant in comparison.

KARL MARX

CR

On its face the Conservation Movement is material, yet in truth there has never been in all human history a popular movement more firmly grounded in ethics, in eternal verities, in the divinity of human rights.

W. J. MCGEE

CR

Western society has accepted as unquestionable a technological imperative that is quite as arbitrary as the most primitive taboo: not merely the duty to foster invention and constantly to create technological novelties, but equally the duty to surrender to these novelties unconditionally, just because they are offered, without respect to their human consequences.

LEWIS MUMFORD

CR

The use of solar energy has not been opened up because the oil industry does not own the sun.

RALPH NADER

❧

We do not inherit the earth from our ancestors, we borrow it from our children.

NATIVE AMERICAN proverb

❧

You must teach your children that the ground beneath their feet is the ashes of your grandfathers. So that they will respect the land, tell your children that the earth is rich with the lives of our kin. Teach your children what we have taught our children, that the earth is our mother. Whatever befalls the earth befalls the sons of the earth. If men spit upon the ground, they spit upon themselves.

NATIVE AMERICAN wisdom

❧

It is only in a human services society which is labor intensive, rather than capital intensive, that the resources of the Earth will be conserved and human resources be expended for the benefit of human beings.

ARTHUR PEARL

❧

World-wide practice of conservation and the fair and continued access by all nations to the resources they need are the two indispensable foundations of continuous plenty and of permanent peace.

Conservation means the wise use of the earth and its resources for the lasting good of men.

The purpose of conservation: The greatest good to the greatest number of people for the longest time.

GIFFORD PINCHOT

℃ℛ

I have no doubt that we will be successful in harnessing the sun's energy. . . . If sunbeams were weapons of war, we would have had solar energy centuries ago.

SIR GEORGE PORTER

℃ℛ

I think the environment should be put in the category of our national security. Defense of our resources is just as important as defense abroad. Otherwise what is there to defend?

ROBERT REDFORD

℃ℛ

Conservation is a great moral issue, for it involves the patriotic duty of insuring the safety and continuance of the nation.

We of an older generation can get along with what we have, though with growing hardship; but in your full manhood and womanhood you will want what nature once so bountifully supplied and man so thoughtlessly destroyed; and because of that want you will reproach us, not for what we have used, but for what we have wasted. . . . So any nation which in its youth lives only for the day, reaps without sowing, and consumes without husbanding, must expect the penalty of the prodigal whose labor could with difficulty find him the bare means of life.

Defenders of the short-sighted men who in their greed and selfishness will, if permitted, rob our country of half its charm by their reckless extermination of all useful and beautiful wild things, sometimes seek to champion them by saying the "game belongs to the people." So it does; and not merely to the people now alive, but to the unborn people. The "greatest good for the greatest number" applies to the number within the womb of time, compared to which those now alive form but an insignificant fraction. Our duty to the whole, including the unborn generations, bids us restrain an unprincipled present-day minority from wasting the heritage of these unborn generations. The movement for the conservation of wild life and the larger movement for the conservation of all our natural resources are essentially democratic in spirit, purpose, and method.

I recognize the right and duty of this generation to develop and use our natural resources, but I do not recognize the right to waste them, or to rob by wasteful use, the generations that come after us.

The conservation of natural resources is the fundamental problem. Unless we solve that problem it will avail us little to solve all others.

Leave it as it is. The ages have been at work on it and man can only mar it.

To waste, to destroy our natural resources, to skin and exhaust the land instead of using it so as to increase its useful-ness, will result in undermining in the days of our children the very prosperity which we ought by right to hand down to them amplified and developed.

The nation behaves well if it treats the natural resources as assets which it must turn over to the next generation increased, and not impaired in value.

THEODORE ROOSEVELT

CR

You forget that the fruits belong to all and that the land belongs to no one.

JEAN-JACQUES ROUSSEAU

CR

Nature provides a free lunch, but only if we control our appetites.

WILLIAM RUCKELSHAUS

CR

In the end, our society will be defined not only by what we create, but by what we refuse to destroy.

JOHN C. SAWHILL

CR

The only way conservation can work is if it is seen as just part of the fabric of development—part of the fabric of growth of human society.

PETER SELIGMANN

CR

So bleak is the picture . . . that the bulldozer and not the atomic bomb may turn out to be the most destructive invention of the twentieth century.

PHILIP SHABECOFF

CR

What is the nature of a species that knowingly and without good reason exterminates another?

GEORGE SMALL

CR

We are the most dangerous species of life on the planet, and every other species, even the earth itself, has cause to fear our power to exterminate. But we are also the only species which, when it chooses to do so, will go to great effort to save what it might destroy.

WALLACE STEGNER

ରେ

The long fight to save wild beauty represents democracy at its best. It requires citizens to practice the hardest of virtues—self-restraint.

EDWIN WAY TEALE

ରେ

Every creature is better alive than dead, men and moose and pine trees, and he who understands it aright will rather preserve its life than destroy it.

If a man walks in the woods for love of them half of each day, he is in danger of being regarded as a loafer. But if he spends his days as a speculator, shearing off those woods and making the earth bald before her time, he is deemed an industrious and enterprising citizen.

HENRY DAVID THOREAU

ରେ

Because we don't think about future generations, they will never forget us.

HENRIK TIKKANEN

ରେ

How to be green? Many people have asked us this important question. It's really very simple and requires no expert knowledge or complex skills. Here's the answer. Consume less. Share more. Enjoy life.

DEREK WALL

CR

I think having land and not ruining it is the most beautiful art that anybody could ever want to own.

ANDY WARHOL

CR

They kill good trees to put out bad newspapers.

JAMES G. WATT

CR

It's obvious that the key problem facing humanity in the coming century is how to bring a better quality of life—for eight billion or more people—without wrecking the environment entirely in the attempt.

In the end . . . success or failure will come down to an ethical decision, one on which those now living will be judged for generations to come.

EDWARD O. WILSON

The Earth

Civilization . . . wrecks the planet from seafloor to
stratosphere.

RICHARD BACH

∞

No beast has ever conquered the earth; and the natural
world has never been conquered by muscular force.

LIBERTY HYDE BAILEY

∞

To cherish what remains of the Earth and to foster its
renewal is our only legitimate hope of survival.

WENDELL BERRY

∞

To see a world in a grain of sand,
And a heaven in a wild flower,
Hold infinity in the palm of your hand,
And eternity in an hour.

WILLIAM BLAKE

∞

The ultimate test of a moral society is the kind of world
that it leaves to its children.

DIETRICH BONHOEFFER

CR

Our generation has inherited an incredibly beautiful world from our parents and they from their parents. It is in our hands whether our children and their children inherit the same world.

RICHARD BRANSON

CR

The land is sacred. These words are at the core of your being. The land is our mother, the rivers our blood. Take our land away and we die. That is, the Indian in us dies.

MARY BRAVE BIRD

CR

I feel no need for any other faith than my faith in the kindness of human beings. I am so absorbed in the wonder of earth and the life upon it that I cannot think of heaven and angels.

I am comforted by life's stability, by earth's unchangeableness. What has seemed new and frightening assumes its place in the unfolding of knowledge. It is good to know our universe. What is new is only new to us.

PEARL S. BUCK

CR

Our children may save us if they are taught to care properly for the planet; but if not, it may be back to the Ice Age or the caves from where we first emerged. Then we'll have to view the universe above from a cold, dark place. No more jet skis, nuclear weapons, plastic crap, broken pay phones, drugs, cars, waffle irons, or television. Come to think of it, that might not be a bad idea.

JIMMY BUFFETT

႙

I have nothing against the planet per se. I root for the big comet or asteroid as a way of cleansing the planet. The comet or asteroid 65 million years ago is probably what gave us our opening to replace the reptiles.

GEORGE CARLIN

႙

Friend, hast thou considered the "rugged, all-nourishing earth," as Sophocles well names her; how she feeds the sparrow on the housetop, much more her darling man?

THOMAS CARLYLE

႙

Those who contemplate the beauty of the earth find reserves of strength that will endure as long as life lasts.

It is a wholesome and necessary thing for us to turn again to the earth and in the contemplation of her beauties to know of wonder and humility.

Those who dwell, as scientists or laymen, among the beauties and mysteries of the earth are never alone or weary of life.

RACHEL CARSON

႙

There are no boundaries in the real Planet Earth. No United States, no Soviet Union, no China, no Taiwan. . . . Rivers flow unimpeded across the swaths of continents. The persistent tides—the pulse of the sea—do not discriminate; they push against all the varied shores on Earth.

JACQUES COUSTEAU

႙

The world is mud-luscious and puddle-wonderful.

E.E. CUMMINGS

ॐ

There is something fundamentally wrong with treating the earth as if it were a business in liquidation.

HERMAN DALY

ॐ

I pledge to live, work and act in a loving, respectful way towards this Earth that I call home, and towards all who live upon it, every insect, animal, fish, bird, plant, human and tree.

GUY DAUNCEY

ॐ

I am the earth. You are the Earth. The Earth is dying. You and I are murderers.

YMBER DELECTO

ॐ

Love the earth as you would love yourself.

JOHN DENVER

ॐ

I hope to be remembered as someone who made the earth a little more beautiful.

WILLIAM O. DOUGLAS

ॐ

The belief that we can manage the Earth and improve on Nature is probably the ultimate expression of human conceit, but it has deep roots in the past and is almost universal.

RENE DUBOS

છ

Humanity is on the march, earth itself is left behind.
                                                    DAVID EHRENFELD

છ

The sky is the daily bread of the eyes.
                                              RALPH WALDO EMERSON

છ

This is a beautiful planet and not at all fragile. Earth can
withstand significant volcanic eruptions, tectonic cataclysms,
and ice ages. But this canny, intelligent, prolific, and
extremely self-centered human creature has proven himself
capable of more destruction of life than Mother Nature
herself. . . . We've got to be stopped.
                                                   MICHAEL L. FISCHER

છ

The earth will continue to regenerate its life sources only
as long as we and all the peoples of the world do our part to
conserve its natural resources. It is a responsibility which every
human being shares. Through voluntary action, each of us can
join in building a productive land in harmony with nature.
                                                      GERALD R. FORD

છ

Now there is one outstandingly important fact regarding
Spaceship Earth, and that is that no instruction book came
with it.

We are not going to be able to operate our Spaceship
Earth successfully nor for much longer unless we see it as a
whole spaceship and our fate as common. It has to be
everybody or nobody.

                                              R. BUCKMINSTER FULLER

CR

Earth provides enough to satisfy every man's need, but not every man's greed.

To forget how to dig the earth and to tend the soil is to forget ourselves.

MOHANDAS GANDHI

CR

And forget not that the Earth delights to feel your bare feet and the winds long to play with your hair.

KAHLIL GIBRAN

CR

As you sit on the hillside, or lie prone under the trees of the forest, or sprawl wet-legged by a mountain stream, the great door, that does not look like a door, opens.

STEPHEN GRAHAM

CR

We know ourselves to be made from this earth. We know this earth is made from our bodies. For we see ourselves and we are nature. We are nature, seeing nature. Nature weeping. Nature speaking of nature to nature.

SUSAN GRIFFIN

CR

The poetry of the earth is never dead.

> To one who has been long in city spent,
> 'Tis very sweet to look into the fair
> And open face of heaven,—to breathe a prayer
> Full in the smile of the blue firmament.

JOHN KEATS

CR

Man must feel the earth to know himself and recognize
his values. . . . God made life simple. It is man who
complicates it.

CHARLES A. LINDBERGH

CR

The earth we abuse and the living things we kill will, in
the end, take their revenge; for in exploiting their presence
we are diminishing our future.

MARYA MANNES

CR

Man is a blind, witless, low brow, anthropocentric clod
who inflicts lesions upon the earth.

IAN MCHARG

CR

There are no passengers on Spaceship Earth. We are all
crew.

MARSHALL MCLUHAN

CR

There is hope if people will begin to awaken that spiritual
part of themselves, that heartfelt knowledge that we are
caretakers of this planet.

BROOKE MEDICINE EAGLE

CR

I conjure you, my brethren, to remain faithful to earth, and
do not believe those who speak unto you of super-terrestrial
hopes! Poisoners they are, whether they know it or not.

FRIEDRICH NIETZSCHE

CR

When we heal the earth, we heal ourselves.

DAVID ORR

CR

The earth and its resources belong of right to its people.

GIFFORD PINCHOT

CR

It is this earth that, like a kind mother, receives us at our birth, and sustains us when born; it is this alone, of all the elements around us, that is never found an enemy of man.

The waters deluge man with rain, oppress him with hail, and drown him with inundations; the air rushes in storms, prepares the tempest, or lights up the volcano; but the earth, gentle and indulgent, ever subservient to the wants of man, spreads his walks with flowers and his table with plenty; returns with interest every good committed to her care, and though she produces the poison, she still supplies the antidote; though constantly teased more to furnish the luxuries of man than his necessities, yet, even to the last, she continues her kind indulgence, and when life is over she piously covers his remains in her bosom.

PLINY THE ELDER

CR

Man has lost the capacity to foresee and to forestall. He will end by destroying the earth.

ALBERT SCHWEITZER

CR

The earth, that is nature's mother, is her tomb.

WILLIAM SHAKESPEARE

❧

Speak no harsh words of earth; she is our mother, and few of us her sons who have not added a wrinkle to her brow.

ALEXANDER SMITH

❧

What humbugs we are, who pretend to live for Beauty, and never see the Dawn!

LOGAN PEARSALL SMITH

❧

Sunshine has no budget, the sea no red tape.

JAREB TEAGUE

❧

No generation has a free hold on this earth. All we have is a life tenancy—with a full repairing lease.

MARGARET THATCHER

❧

Shall I not have intelligence with the earth? Am I not partly leaves and vegetable mould myself.

What's the use of a fine house if you haven't got a tolerable planet to put it on?

HENRY DAVID THOREAU

❧

Don't blow it—good planets are hard to find.

TIME magazine

❧

We are pilgrims, not settlers; this earth is our inn, not our home.

JOHN H. VINCENT

CR

We could have saved the Earth but we were too damned cheap.

KURT VONNEGUT

CR

Earth, thou great footstool of our God
Who reigns on high; thou fruitful source
Of all our raiment, life and food,
Our house, our parent, and our nurse.

ISAAC WATTS

CR

The materials of wealth are in the earth, in the seas, and in their natural and unaided productions.

DANIEL WEBSTER

CR

I am pessimistic about the human race because it is too ingenious for its own good. Our approach to nature is to beat it into submission. We would stand a better chance of survival if we accommodated ourselves to this planet and viewed it appreciatively instead of skeptically and dictatorially.

E. B. WHITE

CR

One does not sell the Earth upon which the people walk.

TASHUNKA WITKO ("CRAZY HORSE")

CR

The common growth of Mother Earth
Suffices me—her tears, her mirth,
Her humblest mirth and tears.

WILLIAM WORDSWORTH

ভ

Where is the dust that has not been alive?
The spade, the plough, disturb our ancestors;
From human mould we reap our daily bread.

EDWARD YOUNG

Ecology

Growth for the sake of growth is the ideology of the cancer cell.

EDWARD ABBEY

❧

Let us be good stewards of the Earth we inherited. All of us have to share the Earth's fragile ecosystems and precious resources, and each of us has a role to play in preserving them. If we are to go on living together on this earth, we must all be responsible for it.

KOFI ANNAN

❧

The natural world is the larger sacred community to which we belong. To be alienated from this community is to become destitute in all that makes us human. To damage this community is to diminish our own existence.

THOMAS BERRY

❧

With laissez-faire and price atomic,
Ecology's Uneconomic,
But with another kind of logic
Economy's Unecologic.

31

Are we to regard the world of nature simply as a storehouse to be robbed for the immediate benefit of man? . . . Does man have any responsibility for the preservation of a decent balance in nature, for the preservation of rare species, or even for the indefinite continuance of his race?

KENNETH E. BOULDING

CR

In America today you can murder land for private profit. You can leave the corpse for all to see, and nobody calls the cops.

PAUL BROOKS

CR

To find the universal elements enough; to find the air and the water exhilarating; to be refreshed by a morning walk or an evening saunter; to be thrilled by the stars at night; to be elated over a bird's nest or a wildflower in spring—these are some of the rewards of the simple life.

How beautiful the leaves grow old. How full of light and color are their last days.

JOHN BURROUGHS

CR

Only within the moment of time represented by the present century has one species—man—acquired significant power to alter the nature of his world.

RACHEL CARSON

CR

Organic farming has been shown to provide major benefits for wildlife and the wider environment. The best that can be

said about genetically engineered crops is that they will now
be monitored to see how much damage they cause.

PRINCE CHARLES

CR

You know why there are so many whitefish in the
Yellowstone River? Because the Fish and Game people have
never done anything to help them.

RUSSELL CHATHAM

CR

Everything is connected to everything else. Everything
must go somewhere. Nature knows best. There is no such
thing as a free lunch. If you don't put something in the
ecology, it's not there.

BARRY COMMONER

CR

We have to shift our emphasis from economic efficiency
and materialism towards a sustainable quality of life and to
healing of our society, of our people and our ecological
systems.

JANET HOLMES À COURT

CR

Is it too late to prevent us from self-destructing? No, for
we have the capacity to design our own future, to take a
lesson from living things around us and bring our values and
actions in line with ecological necessity. But we must first
realize that ecological and social and economic issues are all
deeply intertwined. There can be no solution to one without
a solution to the others.

JEAN-MICHEL COUSTEAU

CR

When the soil disappears, the soul disappears.

YMBER DELECTO

ભ

Man will survive as a species for one reason: He can adapt to the destructive effects of our power-intoxicated technology and of our ungoverned population growth, to the dirt, pollution and noise of a New York or Tokyo. And that is the tragedy. It is not man the ecological crisis threatens to destroy but the quality of human life.

Man shapes himself through the decisions that shape his environment.

RENE DUBOS

ભ

The great ecosystems are like complex tapestries—a million complicated threads, interwoven, make up the whole picture. Nature can cope with small rents in the fabric; it can even, after a time, cope with major disasters like floods, fires, and earthquakes. What nature cannot cope with is the steady undermining of its fabric by the activities of man.

GERALD DURRELL

ભ

If we have a hope of really understanding our place in nature and of carving out a place for ourselves that is sustainable, it's primarily because of the new level of communication. It used to be, "What you don't have in your mind, you have on your shelf." But now we have the Web.

We are all together in this, we are all together in this single living ecosystem called planet earth. As we learn how we fit into the greater scheme of things, and begin to understand how the system works, we can plan ahead, we can use the resources responsibly, to show some respect for this inheritance that goes back 4.6 billion years.

We are in trouble now, unless we deliberately take actions to take care of the sea, and make sure these systems continue to operate as they have for millions of years.

SYLVIA EARLE

CR

Modern technology
Owes ecology
An apology.

ALAN M. EDDISON

CR

We consider species to be like a brick in the foundation of a building. You can probably lose one or two or a dozen bricks and still have a standing house. But by the time you've lost 20 percent of species, you're going to destabilize the entire structure. That's the way ecosystems work.

DONALD FALK

CR

When we protect the places where the processes of life can flourish, we strengthen not only the future of medicine, agriculture and industry, but also the essential conditions for peace and prosperity.

HARRISON FORD

CR

Our environmental problems originate in the hubris of imagining ourselves as the central nervous system or the brain of nature. We're not the brain, we are a cancer on nature.

DAVE FOREMAN

❧

Understanding the laws of nature does not mean that we are immune to their operations.

DAVID GERROLD

❧

You cannot get through a single day without having an impact on the world around you. What you do makes a difference, and you have to decide what kind of difference you want to make.

JANE GOODALL

❧

I think our main foreign policy is economic policy, but to think that economic policy is not environmental policy is to, sort of, miss the point. You know you can't have economic development without impact on the biosphere.

RANDY HAYES

❧

The command "Be fruitful and multiply" was promulgated, according to our authorities, when the population of the world consisted of two people.

WILLIAM RALPH INGE

❧

For if one link in nature's chain might be lost, another might be lost, until the whole of things will vanish by piecemeal.

THOMAS JEFFERSON

❧

Modern society will find no solution to the ecological problem unless it takes a serious look at its lifestyles.

Equally worrying is the ecological question which accompanies the problem of consumerism and which is closely connected to it. In his desire to have and to enjoy rather than to be and to grow, man consumes the resources of the earth and his own life in an excessive and disordered way. At the root of the senseless destruction of the natural environment lies an anthropological error, which unfortunately is widespread in our day. Man, who discovers his capacity to transform and in a certain sense create the world through his own work, forgets that this is always based on God's prior and original gift of the things that are. Man thinks that he can make arbitrary use of the earth, subjecting it without restraint to his will, as though it did not have its own requisites and a prior God-given purpose, which man can indeed develop but must not betray. Instead of carrying out his role as a cooperator with God in the work of creation, man sets himself up in place of God and thus ends up provoking a rebellion on the part of nature, which is more tyrannized than governed by him.

POPE JOHN PAUL II

☙

The range of human and ecological processes that are inherent in each human-environment relationship . . . illustrate how the "thickness" of historical description increases as additional processes are incorporated into theory. . . . If the theories that historians employ help separate the strands of change into identifiable threads, their ability to trace change over time with precision will be increased. The choice among which of those histories is the "correct" interpretation is far from a flight into relativity; rather it is a matter of conscious social decision-making that is itself subject to change over time.

BARBARA LEIBHARDT

❦

In June as many as a dozen species may burst their buds on a single day. No man can heed all of these anniversaries; no man can ignore all of them.

ALDO LEOPOLD

❦

Yet, despite our many advances, our environment is still threatened by a range of problems, including global climate change, energy dependence on unsustainable fossil fuels, and loss of biodiversity.

DAN LIPINSKI

❦

Natural species are the library from which genetic engineers can work.

THOMAS E. LOVEJOY

❦

Any explanation of environmental change should account for the mutually constitutive nature of ecology, production, and cognition, the latter at the level of individuals, which we call ideology, or at the societal level, which in the modern world we call law. . . . To externalize any of the three elements . . . is to miss the crucial fact that human life and thought are embedded in each other and together in the non-human world.

ARTHUR MCEVOY

❦

Within the various acts of the ecodrama should be included scenes in which men's and women's roles come to center stage and scenes in which Nature "herself" is an actress.

CAROLYN MERCHANT

❧

If we human beings learn to see the intricacies that bind one part of a natural system to another and then to us, we will no longer argue about the importance of wilderness protection, or over the question of saving endangered species, or how human communities must base their economic futures—not on short-term exploitation—but on long-term, sustainable development.

If we learn, finally, that what we need to manage is not the land so much as ourselves in the land, we will have turned the history of American land-use on its head.

GAYLORD NELSON

❧

The value of biodiversity is more than the sum of its parts.

BYRAN G. NORTON

❧

The least movement is of importance to all nature. The entire ocean is affected by a pebble.

Nature is an infinite sphere of which the center is everywhere and the circumference nowhere.

BLAISE PASCAL

❧

We're finally going to get the bill for the Industrial Age. If the projections are right, it's going to be a big one: the ecological collapse of the planet.

JEREMY RIFKIN

❧

The system of nature, of which man is a part, tends to be

self-balancing, self-adjusting, self-cleansing. Not so with technology.

E. F. Schumacher

☙

Humankind has not woven the web of life. We are but one thread within it. Whatever we do to the web, we do to ourselves. All things are bound together. All things connect.

Chief Seattle

☙

Guns have metamorphosed into cameras in this earnest comedy, the ecology safari, because nature has ceased to be what it always had been—what people needed protection from. Now nature—tamed, endangered, mortal—needs to be protected from people.

Susan Sontag

☙

We have been god-like in the planned breeding of our domesticated plants, but rabbit-like in the unplanned breeding of ourselves.

Arnold Toynbee

☙

We cannot cheat on DNA. We cannot get round photo-synthesis. We cannot say I am not going to give a damn about phytoplankton. All these tiny mechanisms provide the pre-conditions of our planetary life. To say we do not care is to say in the most literal sense that "we choose death."

Dame Barbara Ward

☙

Why do people give each other flowers? To celebrate

various important occasions, they're killing living creatures?
Why restrict it to plants? Sweetheart, let's make up. Have this
deceased squirrel.

The WASHINGTON POST

CR

Imperialism engenders a particular type of ecological
drama involving several characteristic phases or acts. The play
has been repeated many times, and as with all classical drama,
the plot is now well understood. Indeed some might argue
that there is a depressing repetitiveness to the successive enact-
ments of the colonial eco-drama, as if man and nature knew
how to write only one scenario and insisted upon staging the
same play in theater after theater on an ever-expanding
worldwide tour.

TIMOTHY WEISKEL

CR

We shall continue to have a worsening ecologic crisis until
we reject the Christian axiom that nature has no reason for
existence save to serve man.

LYNN WHITE, JR.

CR

You must not know too much, or be too precise or
scientific about birds and trees and flowers and water-craft; a
certain free margin, and even vagueness—perhaps ignorance,
credulity—helps your enjoyment of these things.

WALT WHITMAN

CR

Nature reserves the right to inflict upon her children the
most terrifying jests.

THORNTON WILDER

❦

Without habitat, there is no wildlife. It's that simple.

WILDLIFE HABITAT CANADA

❦

In its broadest ecological context, economic development is the development of more intensive ways of exploiting the natural environment.

RICHARD WILKINSON

❦

If all mankind were to disappear, the world would regenerate back to the rich state of equilibrium that existed ten thousand years ago. If insects were to vanish, the environment would collapse into chaos.

Humanity is part of nature, a species that evolved among other species. The more closely we identify ourselves with the rest of life, the more quickly we will be able to discover the sources of human sensibility and acquire the knowledge on which an enduring ethic, a sense of preferred direction, can be built.

EDWARD O. WILSON

❦

We may be entering a new phase of history, a time when we begin to rediscover . . . the traditional teaching that power must entail restraint and responsibility, the ancient awareness that we are interdependent with all of nature and that our sense of community must take in the whole of creation.

DONALD WORSTER

❦

The goal of life is living in agreement with nature.

ZENO

Environmentalism

It is horrifying that we have to fight our own government to save our environment.

ANSEL ADAMS

CR

That which is not good for the bee-hive cannot be good for the bees.

MARCUS AURELIUS

CR

The major problems in the world are the result of the difference between how nature works and the way people think.

GREGORY BATESON

CR

In the long term, the economy and the environment are the same thing. If it's unenvironmental it is uneconomical. That is the rule of nature.

MOLLIE BEATTIE

CR

This is a deeply spiritual issue . . . Do we want to spend more time trying to care for our fellow man or do we want

43

to just pursue more virtual reality? That's the issue before us . . . and it's being played out in the world of the environment.

Kids can help the environment by riding a bike. Always wear a helmet of course and stay in the bike lane. Take public transportation with your parents and your friends and see if you like that. That's a good way to get around. Start a home garden, be energy efficient, turn off the lights and the water. All of those things are very good for the environment and good for your pocketbook.

ED BEGLEY, JR.

☙

Can we actually suppose that we are wasting, polluting, and making ugly this beautiful land for the sake of patriotism and the love of God? Perhaps some of us would like to think so, but in fact this destruction is taking place because we have allowed ourselves to believe, and to live, a mated pair of economic lies: that nothing has a value that is not assigned to it by the market; and that the economic life of our communities can safely be handed over to the great corporations.

WENDELL BERRY

☙

If we go on as we are, we will destroy in the next century everything that the poets have been singing about for the past two thousand years.

FRED BODSWORTH

☙

The role of the market place is to be an instrument of environmental change and policy making. We are all

consumers with a great potential for change. Environmental
protection begins at home.

NOEL BROWN

ॐ

I do not know of any environmental group in any country
that does not view its government as an adversary.

GRO HARLEM BRUNDTLAND

ॐ

Energy will be the immediate test of our ability to unite
this nation, and it can also be the standard around which we
rally. On the battlefield of energy we can win for our nation a
new confidence, and we can seize control again of our
common destiny.

Acknowledging the physical realities of our planet does
not mean a dismal future of endless sacrifice. In fact,
acknowledging these realities is the first step in dealing with
them. We can meet the resource problems of the world—
water, food, minerals, farmlands, forests, overpopulation,
pollution—if we tackle them with courage and foresight.

We can drift along as though there were still a cold war,
wasting hundreds of billions of dollars on weapons that will
never be used, ignoring the problems of people in this country
and around the world, being one of the worst environmental
violators on earth, standing against any sort of viable programs
to protect the world's forests or to cut down on acid rain or
the global warming or ozone depletion. We can ignore
human rights violations in other countries, or we can take
these things on as true leaders ought to and accept the
inspiring challenge of America for the future.

JIMMY CARTER

ॐ

Man has been endowed with reason, with the power to create, so that he can add to what he's been given. But up to now he hasn't been a creator, only a destroyer. Forests keep disappearing, rivers dry up, wildlife's become extinct, the climate's ruined and the land grows poorer and uglier every day.

ANTON CHEKHOV

CR

I'm not an environmentalist. I'm an Earth warrior.

DARRYL CHERNEY

CR

Economic advance is not the same thing as human progress.

JOHN CLAPHAM

CR

I have never believed we had to choose between either a clean and safe environment or a growing economy. Protecting the health and safety of all Americans doesn't have to come at the expense of our economy's bottom line. And creating thriving companies and new jobs doesn't have to come at the expense of the air we breathe, the water we drink, the food we eat, or the natural landscape in which we live. We can, and indeed must, have both.

BILL CLINTON

CR

Most of us have become Ecozombies, desensitized, environmental deadheads. On average, society conditions us to spend over 95 percent of our time and 99.9 percent of our thinking disconnected from nature. Nature's extreme absence in our lives leaves us abandoned and wanting. We feel we never have enough. We greedily, destructively, consume and can't stop. Nature's loss in our psyche produces a hurt, hungering void within us that bullies us into our dilemmas.

MICHAEL J. COHEN

$\boxed{\text{CR}}$

It is our collective and individual responsibility to protect and nurture the global family, to support its weaker members and to preserve and tend to the environment in which we all live.

DALAI LAMA

$\boxed{\text{CR}}$

Today's world is one in which the age-old risks of human-kind—the drought, floods, communicable diseases—are less of a problem than ever before. They have been replaced by risks of humanity's own making—the unintended side-effects of beneficial technologies and the intended effects of the technologies of war. Society must hope that the world's ability to assess and manage risks will keep pace with its ability to create them.

J. CLARENCE DAVIES

$\boxed{\text{CR}}$

No one person has to do it all but if each one of us follow our heart and our own inclinations we will find the small things that we can do to create a sustainable future and a healthy environment.

JOHN DENVER

$\boxed{\text{CR}}$

The concept of the public welfare is broad and inclusive . . . the values it represents are spiritual as well as physical, aesthetic as well as monetary. It is within the power of the legislature to determine that the community should be

beautiful as well as healthy, spacious as well as clean, well
balanced as well as carefully patrolled.

WILLIAM O. DOUGLAS

೧೪

It's not easy being a green conservative, but if we conser-
vatives want to be true to our principles we have to move in
that direction. It is morally right. It is religiously correct. It is
economically prudent. It strengthens national defense. And it
makes a better world for our children, and our children's
children.

ROD DREHER

೧೪

It is every man's obligation to put back into the world at
least the equivalent of what he takes out of it.

ALBERT EINSTEIN

೧೪

We must make this an insecure and inhospitable place for
capitalists and their projects. . . . We must reclaim the roads
and plowed land, halt dam construction, tear down existing
dams, free shackled rivers and return to wilderness millions of
tens of millions of acres of presently settled land.

DAVID FOREMAN

೧೪

We need a new environmental consciousness on a global
basis. To do this, we need to educate people.

MIKHAIL GORBACHEV

೧೪

The struggle to save the global environment is in one way
much more difficult than the struggle to vanquish Hitler, for

this time the war is with ourselves. We are the enemy, just as we have only ourselves as allies.

AL GORE

∞

Human consciousness arose but a minute before midnight on the geological clock. Yet we mayflies try to bend an ancient world to our purposes, ignorant perhaps of the messages buried in its long history. Let us hope that we are still in the early morning of our April day.

We cannot win this battle to save species and environments without forging an emotional bond between ourselves and nature as well—for we will not fight to save what we do not love.

STEPHEN JAY GOULD

∞

Economic policy turns out to be the most important environmental policy.

RANDY HAYES

∞

I wake up in the morning asking myself what can I do today, how can I help the world today.

JULIA BUTTERFLY HILL

∞

Where the quality of life goes down for the environment, the quality of life goes down for humans.

GEORGE HOLLAND

∞

The environment, after all, is where we all meet, where

we all have a mutual interest. It is one thing that all of us share. It is not only a mirror of ourselves, but a focusing lens on what we can become.

CLAUDIA ALTA (LADY BIRD) JOHNSON

ભ

The control man has secured over nature has far outrun his control over himself.

ERNEST JONES

ભ

Christianity, with its roots in Judaism, was a major factor in the development of the Western worldview. . . . A basic Christian belief was that God gave humans dominion over creation, with the freedom to use the environment as they saw fit. Another important Judeo-Christian belief predicted that God would bring a cataclysmic end to the Earth sometime in the future. One interpretation of this belief is that the Earth is only a temporary way station on the soul's journey to the afterlife. Because these beliefs tended to devalue the natural world, they fostered attitudes and behaviors that had a negative effect on the environment.

We must not be forced to explore the universe in search of a new home because we have made the Earth inhospitable, even uninhabitable. For if we do not solve the environmental and related social problems that beset us on Earth—pollution, toxic contamination, resource depletion, prejudice, poverty, hunger—those problems will surely accompany us to other worlds.

DONALD G. KAUFMAN and CECILIA M. FRANZ

ભ

The supreme reality of our time is . . . the vulnerability of our planet.

It is our task in our time and in our generation, to hand down undiminished to those who come after us, as was handed down to us by those who went before, the natural wealth and beauty which is ours.

JOHN F. KENNEDY

❧

We won't have a society if we destroy the environment.

MARGARET MEAD

❧

The scarcest resource is not oil, metals, clean air, capital, labor, or technology. It is our willingness to listen to each other and learn from each other and to seek the truth rather than seek to be right.

DONELLA MEADOWS

❧

What have we achieved in mowing down mountain ranges, harnessing the energy of mighty rivers, or moving whole populations about like chess pieces, if we ourselves remain the same restless, miserable, frustrated creatures we were before? To call such activity progress is utter delusion. We may succeed in altering the face of the earth until it is unrecognizable even to the Creator, but if we are unaffected wherein lies the meaning?

HENRY MILLER

❧

Environmentalists have long been fond of saying that the sun is the only safe nuclear reactor, situated as it is some ninety-three million miles away.

STEPHANIE MILLS

❧

Light pollution is increasing. Unless something is done, future generations may never see the stars.

SIR PATRICK MOORE

◌

Give a gift to all generations by saving the earth.

CHRISTINA R. NEWMAN

◌

I am I plus my surroundings and if I do not preserve the latter, I do not preserve myself.

JOSÉ ORTEGA Y GASSET

◌

The activist is not the man who says the river is dirty. The activist is the man who cleans up the river.

ROSS PEROT

◌

Racial injustice, war, urban blight, and environmental rape have a common denominator in our exploitative economic system.

CHANNING E. PHILLIPS

◌

We have been saddled in recent years with a notable set of stock characters, or stereotypes. These include the Noble Savage [and] four joyless rapists—the Forest Service, the farmer, the army engineer, and the lumberman—assaulting Mother Nature, while the Sierra Club purveys chastity belts. . . . Historians who regard conservation as past politics might profit by a spell on the sawmill greenchain, or as trail workers for the Park Service to get some grassroots insights.

LAWRENCE RAKESTRAW

෬

The great task for environmental historians is to record and analyze the effects of man's recently achieved control over the natural world. What is needed is a longer-term global, comparative, historical perspective that treats the environment as a meaningful variable.

JOHN RICHARDS

෬

It hurts the spirit, somehow, to read the word environ-ments, when the plural means that there are so many alternatives there to be sorted through, as in a market, and voted on.

LEWIS THOMAS

෬

It appears to be a law that you cannot have a deep sympathy with both man and nature.

HENRY DAVID THOREAU

෬

Modern man's capacity for destruction is quixotic evidence of humanity's capacity for reconstruction. The powerful technological agents we have unleashed against the environ-ment include many of the agents we require for its reconstruction.

GEORGE F. WILL

෬

Perhaps the time has come to cease calling it the environmentalist view, as though it were a lobbying effort outside the mainstream of human activity, and to start calling it the real-world view.

Darwin's dice have rolled badly for Earth. The human species is, in a word, an environmental abnormality. Perhaps a law of evolution is that intelligence usually extinguishes itself.

EDWARD O. WILSON

ର

It is upsetting that many people don't seem to observe what's happening to the environment, what's happening in terms of global warming, the loss of habitats and wild things.

JOANNE WOODWARD

ର

Sustainable development is . . .development that meets the needs of the present without compromising the ability of further generations to meet their own needs.

WORLD COMMISSION ON ENVIRONMENT AND DEVELOPMENT

Global Warming

The problem of climate change is so large that it can't be solved by voluntary individual responses. It requires an economy-wide solution, i.e., one that limits the total carbon intake of the economy.

PETER BARNES

CR

If you asked me to name the three scariest threats facing the human race, I would give the same answer that most people would: nuclear war, global warming and Windows.

DAVE BARRY

CR

The two most abundant forms of power on earth are solar and wind, and they're getting cheaper and cheaper.

ED BEGLEY, JR.

CR

The blunt truth about the politics of climate change is that no country will want to sacrifice its economy in order to meet this challenge, but all economies know that the only sensible long term way of developing is to do it on a sustainable basis.

Global warming is too serious for the world any longer to ignore its danger or split into opposing factions on it.

TONY BLAIR

ભ

This is not a matter of Chicken Little telling us the sky
is falling. The scientific evidence is telling us we have a
problem, a serious problem.

JOHN CHAFFEE

ભ

I don't mean to imply that we are in imminent danger of
being wiped off the face of the earth—at least, not on account
of global warming. But climate change does confront us with
profound new realities. We face these new realities as a
nation, as members of the world community, as consumers, as
producers, and as investors. And unless we do a better job of
adjusting to these new realities, we will pay a heavy price. We
may not suffer the fate of the dinosaurs. But there will be a
toll on our environment and on our economy, and the toll
will rise higher with each new generation.

EILEEN CLAUSSEN

ભ

We are embarked on the most colossal ecological experi-
ment of all time; doubling the concentration in the atmosphere of an entire planet of one of its most important gases;
and we really have little idea of what might happen.

PAUL A. COLINVAUX

ભ

Deeper than temperature and the extinction of the polar
bear is the idea that we all share this beautiful, ailing planet,
Democrats and Republicans alike.

I feel we are so blessed to live in a country where we enjoy
so many rights that other countries cannot even begin to
imagine. However, it terrifies me that we seem to have lost

touch with our connection to the earth. I am concerned that
we have risen to such heights of arrogance in our refusal to
acknowledge that our earth is rapidly changing in ways that
might affect us catastrophically but instead, we hold steadfast
to our belief that nothing can happen to us as a people.

SHERYL CROW

෬

Global warming has melted the polar ice caps, raised the
levels of the oceans and flooded the earth's great cities.
Despite its evident prosperity, New Jersey is scarcely Utopia.

GODFRIED DANNEELS

෬

Can we really be so uncaring as a country to refuse to
accept our responsibility for global warming when the United
States is the largest contributor to this problem? It breaks my
heart to read every day of another environmental law that is
being chipped away, eroded by this Administration, and we're
not being provided with incentives by Bush to cut back on
our shameful waste of energy. My husband, Bruce, wrote in
one PSA that during World War II the government asked
everyone to make sacrifices for the common good. We must
demand this of our Administration today, for we all have an
obligation as citizens of this earth to leave the world a
healthier, cleaner, and better place for our children and future
generations.

BLYTHE DANNER

෬

The fifth revolution will come when we have spent the
stores of coal and oil that have been accumulating in the earth
during hundreds of millions of years. . . . It is to be hoped
that before then other sources of energy will have been
developed. . . . Whether a convenient substitute for the

present fuels is found or not, there can be no doubt that there will have to be a great change in ways of life. This change may justly be called a revolution, but it differs from all the preceding ones in that there is no likelihood of its leading to increases of population, but even perhaps to the reverse.

                                        CHARLES GALTON DARWIN

◌

Our oil-based society depends on non-renewable resources. It requires relentless probing into vast reaches of pristine land, sacrificing vital bioregions, and irreplaceable cultures. The possibility of catastrophic climate change is substantially increased by the 40 million barrels of oil burned every day by vehicles. We must all move shoulder to shoulder in a unified front to show this administration that the true majority of people are willing to vote for a cleaner environment and won't back down.

                                        LEONARDO DICAPRIO

◌

Yes, there is still much about global warming we have to learn and research should continue. But the longer we delay, the more CO^2 will build up in the atmosphere. It stays there a long time. If we wait too long before acting, we will pass a point of no return and lock ourselves into centuries of global warming. We could pass one of those dangerous tipping points that could make life very difficult. It's a risk we shouldn't take.

                                        JIM DIPESO

◌

Humanity, let us say, is like people packed in a automobile which is traveling downhill without lights at a terrific speed and driven by a four-year-old child. The signposts along the way are all marked progress.

LORD DUNSANY

୧

There are many potential policies to reduce greenhouse-gas emissions for which the total benefits outweigh the total costs. For the United States in particular, sound economic analysis shows that there are policy options that would slow climate change without harming American living standards, and these measures may in fact improve U.S. productivity in the longer run.

ECONOMISTS' STATEMENT ON CLIMATE CHANGE, 2007

୧

What changed in the United States with Hurricane Katrina was a feeling that we have entered a period of consequences.

The key will be a new public awareness of how serious is the threat to the global environment. Those who have a vested interest in the status quo will probably continue to be able to stifle any meaningful change until enough citizens . . . are willing to speak out.

There are many who still do not believe that global warming is a problem at all. And it's no wonder: because they are the targets of a massive and well-organized campaign of disinformation lavishly funded by polluters who are determined to prevent any action to reduce the greenhouse gas emissions that cause global warming out of a fear that their profits might be affected if they had to stop dumping so much pollution into the atmosphere.

The scientists are virtually screaming from the rooftops now. The debate is over! There's no longer any debate in the scientific community about this. But the political systems around the world have held this at arm's length because it's an inconvenient truth, because they don't want to accept that it's a moral imperative.

For a long time, the scientists have been telling us global warming increases the temperature of the top layer in the ocean, and that causes the average hurricane to become a lot stronger. So, the fact that the ocean temperatures did go up because of global warming, because of man-made global warming, starting around in the seventies and then we had a string of unusually strong hurricanes outside the boundaries of this multi-decadal cycle that is a real factor; there are scientists who point that out, and they're right, but we're exceeding those boundaries now.

Al Gore

ᕋ

The sunshine that strikes American roads each year contains more energy than all the fossil fuels used by the entire world.

An acre of windy prairie could produce between $4,000 and $10,000 worth of electricity per year—which is far more than the value of the land's crop of corn or wheat.

America has the technology and resources to meet all its energy needs while safeguarding the earth's climate. The urgent question now is, "Do we have the will?" At least one city does, and I'm proud to live in it.

Denis Hayes

ᕋ

The global warming scenario is pretty grim. I'm not sure I like the idea of polar bears under a palm tree.

Lenny Henry

ᕋ

While human-induced global warming is not going to
turn present-day Earth into present-day Mars, global warming
is dire enough that our most distinguished scientists recently
concluded that as many as one million species on the planet
could be extinct by 2050 if affairs do not change.

JAY INSLEE

જી

America has not led but fled on the issue of global
warming.

JOHN KERRY

જી

The danger posed by war to all of humanity—and to our
planet—is at least matched by the climate crisis and global
warming. I believe that the world has reached a critical stage
in its efforts to exercise responsible environmental stewardship.

BAN KI-MOON

જી

Climate change is the most severe problem that we are
facing today, more serious even than the threat of terrorism.

DAVID KING

જી

People tend to focus on the here and now. The problem
is that, once global warming is something that most people
can feel in the course of their daily lives, it will be too late to
prevent much larger, potentially catastrophic changes.

As best as can be determined, the world is now warmer
than it has been at any point in the last two millennia, and, if
current trends continue, by the end of the century it will
likely be hotter than at any point in the last two million years.

ELIZABETH KOLBERT

CR

Environmental history unites the oldest themes with the newest in contemporary historiography: the evolution of epidemics and climate, those two factors being integral parts of the human ecosystem; the series of natural calamities aggravated by a lack of foresight . . . the destruction of Nature, caused by soaring population and/or by the predators of industrial overconsumption; nuisances of urban and manufacturing origin, which lead to air or water pollution; human congestion or noise levels in urban areas, in a period of galloping urbanization.

EMMANUEL LE ROY LADURIE

CR

If we do nothing, warming could drive the sugar maple right out of Vermont, which would be a catastrophe for my state . . . Congress can't seem to make up its mind about whether it wants a two-year budget, let alone focus on the effects of global warming which may not occur for several decades. It is our job to worry about the long-term future of American agriculture and forestry.

PATRICK J. LEAHY

CR

Today, we can see with our own eyes what global warming is doing. In that context it becomes truly irresponsible, if not immoral, for us not to do something.

JOE LIEBERMAN

CR

Global warning is causing significant harm to California's environment, economy, agriculture and public health. It is time to hold these companies responsible for their contribution to this crisis.

BILL LOCKYER

CR

On average, global warming is not going to harm the developing world.

BJØRN LOMBORG

CR

Some urge we do nothing because we can't be certain how bad the (climate) problem might become or they presume the worst effects are most likely to occur in our grandchildren's lifetime. I'm a proud conservative, and I reject that kind of live-for-today, "me generation," attitude. It is unworthy of us and incompatible with our reputation as visionaries and problem solvers. Americans have never feared change. We make change work for us.

Our nation has both an obligation and self-interest in facing head-on the serious environmental, economic and national security threat posed by global warming.

JOHN MCCAIN

CR

The greenhouse effect is a more apt name than those who coined it imagined. The carbon dioxide and trace gases act like the panes of glass on a greenhouse—the analogy is accurate. But it's more than that. We have built a greenhouse, a human creation, where once there bloomed a sweet and wild garden.

BILL MCKIBBEN

CR

A critical problem for humans is to avoid arriving inadvertently at a critical threshold that might trigger an abrupt accelerated warming of the climate. . . . Animals today are generally adapted to relatively cool conditions, as were faunas prior to the terminal Mesozoic extinctions. A sudden climatic warming could potentially impose on us conditions comparable to those that terminated a geologic age.

DEWEY M. MCLEAN

CR

The issue of climate change is one that we ignore at our own peril. There may still be disputes about exactly how much we're contributing to the warming of the earth's atmosphere and how much is naturally occurring, but what we can be scientifically certain of is that our continued use of fossil fuels is pushing us to a point of no return. And unless we free ourselves from a dependence on these fossil fuels and chart a new course on energy in this country, we are condemning future generations to global catastrophe.

Today we're seeing that climate change is about more than a few unseasonably mild winters or hot summers. It's about the chain of natural catastrophes and devastating weather patterns that global warming is beginning to set off around the world . . . the frequency and intensity of which are breaking records thousands of years old.

All across the world, in every kind of environment and region known to man, increasingly dangerous weather patterns and devastating storms are abruptly putting an end to the long-running debate over whether or not climate change is real. Not only is it real, it's here, and its effects are giving rise to a frighteningly new global phenomenon: the man-made natural disaster.

BARACK OBAMA

CR

The world's forests need to be seen for what they are . . .
giant global utilities, providing essential services to humanity
on a vast scale. Rainforests store carbon, which is lost to the
atmosphere when they burn, increasing global warming. The
life they support cleans the atmosphere of pollutants and feeds
it with moisture. They help regulate our climate and sustain
the lives of some of the poorest people on this Earth.

PRINCE CHARLES

 CR

The answer to global warming is in the abolition of private
property and production for human need. A socialist world
would place an enormous priority on alternative energy
sources. This is what ecologically-minded socialists have been
exploring for quite some time now.

LOUIS PROYECT

CR

We want a national emissions trading scheme, the Government does not and has rejected one for years. We want to
boost the mandatory renewable energies target, the Government has failed to do that. We want a national demand side
management strategy for the country to reduce electricity
consumption and the Government, up until now, has done
very little on that score.

KEVIN RUDD

CR

In my opinion, forecasts of "the end of the world" or
"nothing to worry about" are the least likely cases, with
almost any scenario in between being more probable.

STEPHEN H. SCHNEIDER

CR

I think that if it has to do with global warming, or if it has to do with raising the minimum wage, or if it has to do with lowering prescription drugs for vulnerable citizens . . . all of those things are people issues, not Democratic issues or Republican issues.

The facts are there that we have created—man has—a self-inflicted wound through global warming.

We simply must do everything we can in our power to slow down global warming before it is too late. The science is clear. The global warming debate is over.

ARNOLD SCHWARZENEGGER

CR

There will be no polar ice by 2060 . . . Somewhere along that path, the polar bear drops out.

LARRY SCHWEIGER

CR

Global climate change is real and we have a limited time to change our behavior or live with the consequences. We can all help by making small changes in our lives to letting our voice be heard by our governing bodies. As has always been the case in this country, if the people demand change, it will come.

KYRA SEDGWICK

CR

This is not a disaster, it is merely a change. The area [Bangladesh] won't have disappeared, it will just be under water. Where you now have cows, you will have fish.

J. R. SPRADLEY

CR

We are upsetting the atmosphere upon which all life depends. In the late '80s when I began to take climate change seriously, we referred to global warming as a "slow motion catastrophe," one we expected to kick in perhaps generations later. Instead, the signs of change have accelerated alarmingly.

DAVID SUZUKI

CR

The Los Angelesization of the planet cannot take place, for in the greenhouse effect nature has her own negative feedback mechanisms for shutting down the furnace of industrial civilization.

WILLIAM IRWIN THOMPSON

CR

I'd say the chances are about 50–50 that humanity will be extinct or nearly extinct within 50 years. Weapons of mass destruction, disease, I mean this global warming is scaring the living daylights out of me.

TED TURNER

CR

It is in the midst of this compromised and complex situation—the reciprocal influences of a changing nature and a changing society—that environmental history must find its home.

RICHARD WHITE

CR

The first proof of global warming may well come from the bleaching of the fragile and highly sensitive coral reef system.

ERNEST H. WILLIAMS

CR

Carbon dioxide, until now an apparently innocuous trace gas in the atmosphere, may be moving rapidly toward a central role as a major threat to the present world order.

GEORGE M. WOODWELL

Nature

For myself I hold no preferences among flowers, so long as they are wild, free, spontaneous. Bricks to all greenhouses! Black thumb and cutworm to the potted plant!

EDWARD ABBEY

☙

The sun is the epitome of benevolence—it is life-giving and warmth-giving and happiness-giving, and to it we owe our thanksgiving.

JESSI LANE ADAMS

☙

Innovative capitalists have tried to rewrite nature, but to no avail.

How many stanzas in the springtime breeze?
How plenty the raindrops? As He doth please.
There is no meter and there is no rhyme,
Yet God's poems always read in perfect time.

ASTRID ALAUDA

☙

I am at two with nature.

WOODY ALLEN

CR

The evidence of Nature is worth more than the arguments of learning.

AMBROSE OF MILAN

CR

Nature is a hanging judge.

ANONYMOUS

CR

If one way be better than another, that you may be sure is Nature's way.

All art, all education, can be merely a supplement to nature.

Nature does nothing without purpose or uselessly.

ARISTOTLE

CR

To sit in the shade on a fine day and look upon verdure is the most perfect refreshment.

JANE AUSTEN

CR

In nature things move violently to their place, and calmly in their place.

Nature is often hidden, sometimes overcome, seldom extinguished.

Nature, to be commanded, must be obeyed.

FRANCIS BACON

❧

When we understand that man is the only animal who must create meaning, who must open a wedge into neutral nature, we already understand the essence of love. Love is the problem of an animal who must find life, create a dialogue with nature in order to experience his own being.

ERNEST BECKER

❧

That we find a crystal or a poppy beautiful means that we are less alone, that we are more deeply inserted into existence than the course of a single life would lead us to believe.

JOHN BERGER

❧

The insufferable arrogance of human beings to think that Nature was made solely for their benefit, as if it was conceivable that the sun had been set afire merely to ripen men's apples and head their cabbages.

CYRANO DE BERGERAC

❧

I am not bound for any public place, but for ground of my own where I have planted vines and orchard trees, and in the heat of the day climbed up into the healing shadow of the woods. Better than any argument is to rise at dawn and pick dew-wet red berries in a cup.

Whether we and our politicians know it or not, Nature is a party to all our deals and decisions, and she has more votes, a longer memory, and a sterner sense of justice than we do.

WENDELL BERRY

❧

Nature is the most thrifty thing in the world; she never wastes anything; she undergoes change, but there's no annihilation, the essence remains—matter is eternal.

THOMAS BINNEY

જ્ર

Nature is just enough; but men and women must comprehend and accept her suggestions.

ANTOINETTE BROWN BLACKWELL

જ્ર

Nature always springs to the surface and manages to show what she is. It is vain to stop or try to drive her back. She breaks through every obstacle, pushes forward, and at last makes for herself a way.

NICOLAS BOILEAU-DESPRÉAUX

જ્ર

You can't be suspicious of a tree, or accuse a bird or a squirrel of subversion or challenge the ideology of a violet.

HAL BORLAND

જ્ર

Nature is the art of God.

THOMAS BROWNE

જ્ર

A flower is an educated weed.

Nature's laws affirm instead of prohibit. If you violate her laws you are your own prosecuting attorney, judge, jury, and hangman.

LUTHER BURBANK

ભ

Knowledge of those unalterable Relations which
Providence has ordained that every thing should bear to every
other . . . To these we should conform in good Earnest; and
not think to force Nature, and the whole Order of her
System, by a Compliance with our Pride, and Folly, to
conform to our artificial Regulations.

EDMUND BURKE

ભ

I go to nature to be soothed and healed, and to have my
senses put in order.

Nature teaches more than she preaches. There are no
sermons in stones. It is easier to get a spark out of a stone than
a moral.

JOHN BURROUGHS

ભ

There is a pleasure in the pathless woods,
There is a rapture on the lonely shore,
There is society, where none intrudes,
By the deep sea, and music in its roar:
I love not man the less, but Nature more.

LORD BYRON

ભ

Like music and art, love of nature is a common language
that can transcend political or social boundaries.

JIMMY CARTER

ભ

I love to think of nature as an unlimited broadcasting

station, through which God speaks to us every hour, if we will only tune in.

Reading about nature is fine, but if a person walks in the woods and listens carefully, he can learn more than what is in books, for they speak with the voice of God.

GEORGE WASHINGTON CARVER

ભ

Human nature is just about the only nature some people experience.

ABIGAIL CHARLESON

ભ

The rose has thorns only for those who would gather it.

Chinese proverb

ભ

I never had any other desire so strong, and so like covetousness, as that . . . I might be master at last of a small house and a large garden, with very moderate conveniences joined to them, and there dedicate the remainder of my life to the culture of them and the study of nature.

ABRAHAM COWLEY

ભ

We go to sanctuaries to remember the things we hold most dear, the things we cherish and love. And then—the great challenge—we return home seeking to enact this wisdom as best we can in our daily lives.

WILLIAM CRONON

ભ

I thank you God for this most amazing day, for the leaping

greenly spirits of trees, and for the blue dream of sky and for everything which is natural, which is infinite, which is yes.

E.E. CUMMINGS

ଓ

How strange that Nature does not knock, and yet does not intrude!

EMILY DICKINSON

ଓ

Nature always strikes back. It takes all the running we can do to remain in the same place.

Human destiny is bound to remain a gamble, because at some unpredictable time and in some unforeseeable manner nature will strike back.

RENE DUBOS

ଓ

Look deep into nature, and then you will understand everything better.

A human being is part of the whole, called by us Universe, a part limited in time and space. He experiences himself, his thoughts and feelings as something separated from the rest—a kind of optical delusion of his consciousness. This delusion is a kind of prison for us, restricting us to our personal desires and to affection for a few persons nearest to us. Our task must be to free ourselves from this prison by widening our circle of compassion to embrace all living creatures and the whole [of] nature in its beauty.

ALBERT EINSTEIN

ଓ

Religion, as distinguished from modern paganism, implies

a life in conformity with nature. It may be observed that the natural life and the supernatural life have a conformity to each other which neither has with the mechanistic life. . . . A wrong attitude towards nature implies, somewhere, a wrong attitude towards God. . . . [We should] struggle to recover the sense of relation to nature and to God.

T. S. Eliot

ଔ

Our life is an apprenticeship to the truth that around every circle another can be drawn; that there is no end in nature, but every end is a beginning, and under every deep a lower deep opens.

How cunningly nature hides every wrinkle of her inconceivable antiquity under roses and violets and morning dew!

Everything in nature contains all the power of nature. Everything is made of one hidden stuff.

Adopt the pace of nature: her secret is patience.

The happiest man is he who learns from nature the lesson of worship.

Nature hates calculators.

In the presence of nature, a wild delight runs through the man, in spite of real sorrows.

To speak truly, few adult persons can see nature. Most persons do not see the sun. At least they have a very superficial seeing. The sun illuminates only the eye of the man, but shines into the eye and heart of the child. The lover of nature is he whose inward and outward senses are still truly adjusted to each other; who has retained the spirit of infancy even into the era of manhood.

Ralph Waldo Emerson

CR

There is no birth in mortal things, and no end in ruinous death. There is only mingling and interchange of parts, and it is this we call "nature."

EMPEDOCLES

CR

The sun, with all those planets revolving around it and dependent upon it, can still ripen a bunch of grapes as if it had nothing else in the universe to do.

GALILEO

CR

I can enjoy society in a room; but out of doors, nature is company enough for me.

WILLIAM HAZLITT

CR

Nature cannot be tricked or cheated. She will give up to you the object of your struggles only after you have paid her price.

NAPOLEON HILL

CR

There is a way that nature speaks, that land speaks. Most of the time we are simply not patient enough, quiet enough, to pay attention to the story.

LINDA HOGAN

CR

On every stem, on every leaf . . . and at the root of everything that grew, was a professional specialist in the shape

of grub, caterpillar, aphis, or other expert, whose business it
was to devour that particular part.

OLIVER WENDELL HOLMES

∝

Drive Nature forth by force, she'll turn and rout
The false refinements that would keep her out.

HORACE

∝

Some people say a man's best friend is the dog. Mine is
Nature.

WARD ELLIOT HOUR

∝

I've always regarded nature as the clothing of God.

ALAN HOVHANESS

∝

Never does nature say one thing and wisdom another.

JUVENAL

∝

Some people worry that artificial intelligence will make us
feel inferior, but then, anybody in his right mind should have
an inferiority complex every time he looks at a flower.

ALAN C. KAY

∝

True conformity to the dictates of nature requires
reverence for the past and solicitude for the future. "Nature"
is not simply the sensation of the passing moment; it is eternal,
though we evanescent men experience only a fragment of it.

We have no right to imperil the happiness of posterity by impudently tinkering with the heritage of humanity.

RUSSELL KIRK

❧

My heart that was rapt away by the wild cherry blossoms—will it return to my body when they scatter?

KOTOMICHI

❧

Cats are intended to teach us that not everything in nature has a function.

JOSEPH WOOD KRUTCH

❧

Nature does not hurry, yet everything is accomplished.

Nature is not benevolent; with ruthless indifference she makes all things serve their purpose.

LAO TZU

❧

Watching clouds roll by
on a sunny day
Who needs church?
Nature is divine.

CARRIE LATET

❧

Art gallery? Who needs it? Look up at the swirling silver-lined clouds in the magnificent blue sky or at the silently blazing stars at midnight. How could indoor art be any more masterfully created than God's museum of nature?

GREY LIVINGSTON

❧

For 200 years we've been conquering Nature. Now we're beating it to death.

TOM MCMILLAN

❧

The richness I achieve comes from Nature, the source of my inspiration.

CLAUDE MONET

❧

Let us permit nature to have her way. She understands her business better than we do.

MICHEL DE MONTAIGNE

❧

Birth, life, and death—each took place on the hidden side of a leaf.

TONI MORRISON

❧

Nature is my medicine.

SARA MOSS-WOLFE

❧

The gross heathenism of civilization has generally destroyed nature, and poetry, and all that is spiritual.

Nature chose for a tool, not the earthquake or lightning to rend and split asunder, not the stormy torrent or eroding rain, but the tender snow-flowers noiselessly falling through unnumbered centuries.

Everybody needs beauty as well as bread, places to play in and pray in, where nature may heal and give strength to body and soul.

When one tugs at a single thing in nature, he finds it attached to the rest of the world.

JOHN MUIR

ॐ

Such is the audacity of man, that he hath learned to counterfeit Nature, yea, and is so bold as to challenge her in her work.

PLINY THE ELDER

ॐ

A margin of life is developed by Nature for all living things—including man. All life forms obey Nature's demands—except man, who has found ways of ignoring them.

EUGENE M. POIROT

ॐ

A lawn is nature under totalitarian rule.

MICHAEL POLLAN

ॐ

A wise man can do no better than to turn from the churches and look up through the airy majesty of the wayside trees with exultation, with resignation, at the unconquerable unimplicated sun.

LLEWELYN POWYS

ॐ

What makes the desert beautiful is that somewhere it hides a well.

ANTOINE DE SAINT-EXUPÉRY

❧

Joy all creatures drink
At nature's bosoms.

FRIEDRICH VON SCHILLER

❧

And this, our life, exempt from public haunt, finds tongues in trees, books in the running brooks, sermons in stones, and good in everything.

WILLIAM SHAKESPEARE

❧

The old Lakota was wise. He knew that man's heart away from nature becomes hard; he knew that lack of respect for growing, living things soon led to lack of respect for humans too.

CHIEF LUTHER STANDING BEAR

❧

Nature is man's teacher. She unfolds her treasures to his search, unseals his eye, illumes his mind, and purifies his heart; an influence breathes from all the sights and sounds of her existence.

ALFRED BILLINGS STREET

❧

Where the most beautiful wild-flowers grow, there man's spirit is fed, and poets grow.

Nature abhors a vacuum, and if I can only walk with sufficient carelessness I am sure to be filled.

Nature will bear the closest inspection. She invites us to lay our eye level with her smallest leaf, and take an insect view of its plain.

I believe that there is a subtle magnetism in Nature, which, if we unconsciously yield to it, will direct us aright.

Nature is full of genius, full of the divinity; so that not a snowflake escapes its fashioning hand.

We can never have enough of nature.

HENRY DAVID THOREAU

CR

If people think that nature is their friend, then they sure don't need an enemy.

KURT VONNEGUT

CR

The true religion, it is said, is service to mankind; but this service seems to take the form of securing for him an unconditional victory over nature. Now this attitude is impious, for, as has been noted, it violates the belief that creation or nature is fundamentally good, that the ultimate reason for its laws is a mystery, and that acts of defiance such as are daily celebrated by the newspapers are subversive of cosmos.

Nature is not something to be fought, conquered and changed according to any human whims. To some extent, of course, it has to be used. But what man should seek in regard to nature is not a complete domination but a modus vivendi— that is, a manner of living together, a coming to terms with something that was here before our time and will be here after it. The important corollary of this doctrine, it seems to me, is that man is not the lord of creation, with an omnipotent will, but a part of creation, with limitations, who ought to observe a decent humility in the face of the inscrutable.

RICHARD WEAVER

❦

I would feel more optimistic about a bright future for man if he spent less time proving that he can outwit Nature and more time tasting her sweetness and respecting her seniority.

E. B. WHITE

❦

I believe a leaf of grass is no less than the journey-work of the stars.

WALT WHITMAN

❦

Nature holds the key to our aesthetic, intellectual, cognitive and even spiritual satisfaction.

EDWARD O. WILSON

❦

Study nature, love nature, stay close to nature. It will never fail you.

I believe in God, only I spell it N-a-t-u-r-e.

FRANK LLOYD WRIGHT

❦

Man maketh a death which Nature never made.

EDWARD YOUNG

Pollution

Our modern industrial economy takes a mountain covered
with trees, lakes, running streams, and transforms it into a
mountain of junk, garbage, slime pits, and debris.

EDWARD ABBEY

&

A fuming smokestack is the perfect symbol of our national
dilemma. On the one hand, it means the jobs and products
we need. On the other, it means pollution.

AMERICAN GAS ASSOCIATION advertisement

&

When someone is chronically ill, the cost of pollution to
him is almost infinite.

ANONYMOUS U.S. CONGRESSIONAL STAFF MEMBER

&

Soon silence will have passed into legend. Man has turned
his back on silence. Day after day he invents machines and
devices that increase noise and distract humanity from the
essence of life, contemplation, meditation . . . tooting, howling, screeching, booming, crashing, whistling, grinding, and
trilling bolster his ego. His anxiety subsides. His inhuman void
spreads monstrously like a gray vegetation.

JEAN ARP

☙

One major, overwhelming reason why we are running out
of water is that we are killing the water we have.

WILLIAM ASHWORTH

☙

It is a sign of our power, and our criminal folly, that we
can pollute the vast ocean and are doing so.

ISAAC ASIMOV

☙

I was a Boy Scout, and scouting gave me an appreciation
for nature and the outdoors that set the stage for my career as
an environmental activist. The other thing would be growing
up in Los Angeles—the smog capital of the world, for quite a
while.

I can't imagine a right more basic than the right to breathe
clean air. We've debated for years how that might be possible.
Now that we know it is, will we have the courage and the
conviction to get there?

ED BEGLEY, JR.

☙

In Kentucky, we're destroying mountains, including their
soils and forests, in order to get at the coal. In other words,
we're destroying a permanent value in order to get at an
almost inconceivably transient value. That coal has a value
only if and when it is burnt. And after it is burnt, it is a
pollutant and a waste—a burden.

WENDELL BERRY

☙

As soils are depleted, human health, vitality and intelligence go with them.

LOUIS BROMFIELD

CR

And Man created the plastic bag and the tin and aluminum can and the cellophane wrapper and the paper plate, and this was good because Man could then take his automobile and buy all his food in one place and He could save that which was good to eat in the refrigerator and throw away that which had no further use. And soon the earth was covered with plastic bags and aluminum cans and paper plates and disposable bottles and there was nowhere to sit down or walk, and Man shook his head and cried: "Look at this God-awful mess."

ART BUCHWALD

CR

We can use our scientific knowledge to improve and beautify the earth, or we can use it to . . . poison the air, corrupt the waters, blacken the face of the country, and harass our souls with loud and discordant noises, [or]. . . we can use it to mitigate or abolish all these things.

JOHN BURROUGHS

CR

Oh Beautiful for smoggy skies, insecticided grain,
For strip-mined mountain's majesty above the asphalt plain.
America, America, man sheds his waste on thee,
And hides the pines with billboard signs, from sea to oily sea.

GEORGE CARLIN

CR

For the first time in the history of the world, every human

being is now subjected to contact with dangerous chemicals, from the moment of conception until death.

No witchcraft, no enemy action had silenced the rebirth of new life in this stricken world. The people had done it themselves.

It is a curious situation that the sea, from which life first arose, should now be threatened by the activities of one form of that life. But the sea, though changed in a sinister way, will continue to exist; the threat is rather to life itself.

RACHEL CARSON

CR

In an underdeveloped country, don't drink the water; in a developed country, don't breathe the air.

CHANGING TIMES magazine

CR

If you could tomorrow morning make water clean in the world, you would have done, in one fell swoop, the best thing you could have done for improving human health by improving environmental quality.

WILLIAM C. CLARK

CR

As we gain satisfaction from artificial substitutes for nature we forget that there is no known substitute for nature, the real thing and its eons of intelligent, life supportive, experience. Each substitute we create falls short of nature's balanced perfection, thus producing our pollution, garbage and relationship conflicts.

MICHAEL J. COHEN

CR

A new generation is being raised—with DDT in their fat, carbon monoxide in their systems and lead in their bones. That is technological man.

As large a body of water as Lake Erie has already been overwhelmed by pollutants and has in effect died.

BARRY COMMONER

ca

The sea is the universal sewer.

JACQUES COUSTEAU

ca

Only when the last tree has died and the last river been poisoned and the last fish been caught will we realize we cannot eat money.

CREE INDIAN proverb

ca

Fresh air keeps the doctor poor.

DANISH proverb

ca

When you defile the pleasant streams
And the wild bird's abiding place,
You massacre a million dreams
And cast your spittle in God's face.

JOHN DRINKWATER

ca

The most important pathological effects of pollution are extremely delayed and indirect.

RENE DUBOS

CR

The automobile has not merely taken over the street, it has dissolved the living tissue of the city . . . Gas-filled, noisy and hazardous, our streets have become the most inhumane landscape in the world.

JAMES M. FITCH

CR

Pollution is nothing but the resources we are not harvesting. We allow them to disperse because we've been ignorant of their value.

R. BUCKMINSTER FULLER

CR

The main lesson to be learned from the Love Canal crisis is that in order to protect public health from chemical contamination, there needs to be a massive outcry—a choir of voices—by the American people demanding change.

The citizens of Love Canal provided an example of how a blue-collar community with few resources can win against great odds (a multi-billion-dollar international corporation and an unresponsive government), using the power of the people in our democratic system.

It will take a massive effort to move society from corporate domination, in which industry's rights to pollute and damage health and the environment supersede the public's right to live, work, and play in safety. This is a political fight. The science is already there, showing that people's health is at risk. To win, we will need to keep building the movement, networking with one another, planning, strategizing, and moving forward. Our children's futures, and those of their unborn children, are at stake.

LOIS GIBBS

❦

It wasn't the Exxon Valdez captain's driving that caused the Alaskan oil spill. It was yours.

GREENPEACE ADVERTISEMENT

❦

The poor shall inherit the earth . . . and all of the toxic waste thereof.

GREENPEACE SLOGAN

❦

Give a man a fish, and he can eat for a day. But teach a man how to fish, and he'll be dead of mercury poisoning inside of three years.

CHARLES HAAS

❦

Acid rain spares nothing. What has taken humankind decades to build and nature millennia to evolve is being impoverished and destroyed in a matter of years—a mere blink in geologic time.

DON HINRICHSEN

❦

Unfortunately, our affluent society has also been an effluent society.

HUBERT H. HUMPHREY

❦

The American people have a right to air that they and their children can breathe without fear.

The air and water grow heavier with the debris of our spectacular civilization.

LYNDON BAINES JOHNSON

❧

The issue is really whether we are going to recognize that the oceans, like the land, are not limitless, self-healing, and invulnerable to humanity's harmful activities.

ELIZABETH KAPLAN

❧

We in government have begun to recognize the critical work which must be done at all levels—local, state and federal—in ending the pollution of our waters.

ROBERT F. KENNEDY

❧

In 22 states, parents can't take kids fishing and eat the fish if they're lucky enough to catch anything because of mercury. Think about that.

JOHN KERRY

❧

The ocean is tired. It's throwing back at us what we're throwing in there.

FRANK LAUTENBERG

❧

The automobile companies are spending too much money on goodwill and too little on trying to see if there is an alternative to the internal combustion engine.

WILLIAM P. LEAR

❧

How long can men thrive between walls of brick, walking on asphalt pavements, breathing the fumes of coal and of oil, growing, working, dying, with hardly a thought of wind, and sky, and fields of grain, seeing only machine-made beauty, the mineral-like quality of life?

CHARLES A. LINDBERGH

ભ

When you use a manual push mower, you're cutting down on pollution and the only thing in danger of running out of gas is you!

GREY LIVINGSTON

ભ

For instance, the average American car driven the average American distance—ten thousand miles—in an average American year releases its own weight in carbon into the atmosphere. Imagine each car on a busy freeway pumping a ton of carbon into the atmosphere, and the sky seems less infinitely blue.

BILL MCKIBBEN

ભ

Pollution doesn't carry a passport.

THOMAS MCMILLAN

ભ

The solution to pollution is local self-reliance.

Pollution often disappears when we switch to renewable resources.

DAVID MORRIS

ભ

To what degree and how often do we have to bruise this delicate living surface of the oceans with oil and other pollutants before the whole system collapses or is destructively and irreversibly diminished?

Every tanker, however well managed, drops some of its oil into the sea in some form or another; badly managed ships are ceaseless polluters and, like garden snails, can often be followed by the long iridescent trail of their waste.

NOEL MOSTERT

CR

The governing classes nowadays want to talk about social justice and projects that provide jobs for lawyers. No one wants to talk about sewers.

DANIEL PATRICK MOYNIHAN

CR

I know that our bodies were made to thrive only in pure air, and the scenes in which pure air is found.

JOHN MUIR

CR

Federal policy over the past century has largely failed to promote an energy system based on safe, secure, economically affordable, and environmentally benign energy sources. The tax code, budget appropriations, and regulatory processes overwhelmingly have been used to subsidize dependence on fossil fuels and nuclear power. The result: increased sickness and premature deaths, depleted family budgets, acid rain, destruction of lakes, forests, and crops, oil spill contamination, polluted rivers and loss of aquatic species and the long-term peril of climate change and radioactive waste dumps—not to mention a dependency on external energy supplies.

Sanctions against polluters are feeble and out of date, and are rarely invoked.

Air pollution (and its fallout on soil and water) is a form of domestic chemical and biological warfare.

Obviously, the answer to oil spills is to paper-train the tankers.

RALPH NADER

ᘓ

Remember when atmospheric contaminants were romantically called stardust?

LANE OLINGHOUSE

ᘓ

There's so much pollution in the air now that if it weren't for our lungs there'd be no place to put it all.

ROBERT ORBEN

ᘓ

We see great damage to specific areas of the seas. They are disease spots like a black spot on the lung.

ARVID PARDO

ᘓ

Dig a trench through a landfill and you will see layers of phone books like geographical strata or layers of cake. . . . During a recent landfill dig in Phoenix, I found newspapers dating from 1952 that looked so fresh you might read one over breakfast.

WILLIAM RATHJE

ᘓ

Trees cause more pollution than automobiles do.

RONALD REAGAN

ભજ

Water has been the orphan stepchild of the entire
conservation picture and our polluted streams have been a
national disgrace for years.

KENNETH REID

ભજ

Pollution is a crime compounded of ignorance and avarice.

LORD RITCHIE-CALDER

ભજ

Here in the United States we turn our rivers and streams
into sewers and dumping grounds, we pollute the air, we
destroy forests, and exterminate fishes, birds, and mammals—
not to speak of vulgarizing charming landscapes with hideous
advertisements. But at last it looks as if our people are
awakening.

THEODORE ROOSEVELT

ભજ

You go into a community and they will vote 80 percent
to 20 percent in favor of a tougher Clean Air Act, but if you
ask them to devote 20 minutes a year to having their car
emissions inspected, they will vote 80 to 20 against it. We
are a long way in this country from taking individual
responsibility for the environmental problem.

WILLIAM RUCKELSHAUS

ભજ

One of the first laws against air pollution came in 1300
when King Edward I decreed the death penalty for burning

of coal. At least one execution for that offense is recorded. But economics triumphed over health considerations, and air pollution became an appalling problem in England.

GLENN T. SEABORG

❧

I durst not laugh for fear of opening my lips and receiving the bad air.

WILLIAM SHAKESPEARE

❧

The new American finds his challenge and his love in the traffic-choked streets, skies nested in smog, choking with the acids of industry, the screech of rubber and houses leashed in against one another while the townlets wither a time and die.

JOHN STEINBECK

❧

Man is a complex being: he makes deserts bloom—and lakes die.

GIL STERN

❧

One can exist for days without food or water or companionship or sex or mental stimulation, but life without air is measured in seconds. In *seconds*.

CASKIE STINNETT

❧

The greed of the rich is often the root cause of serious pollution problems. In such cases we can only pray, "God do not forgive them, for they know what they do."

MONKO SWAMINATHAN

CR

As we watch the sun go down, evening after evening, through the smog across the poisoned waters of our native earth, we must ask ourselves seriously whether we really wish some future universal historian on another planet to say about us: With all their genius and with all their skill, they ran out of foresight and air and food and water and ideas, or, they went on playing politics until their world collapsed around them.

U THANT

CR

Thank God men cannot fly, and lay waste the sky as well as the earth.

HENRY DAVID THOREAU

CR

Pollution is the forerunner of perdition.

JOHN TRAPP

CR

Why should man expect his prayer for mercy to be heard by What is above him when he shows no mercy to what is under him?

PIERRE TROUBETZKOY

CR

We have found the sources of hazardous waste and they are us.

U.S. ENVIRONMENTAL PROTECTION AGENCY

CR

Recommendations that children not run to and from school and that events be suspended are not a substitute for reducing pollution.

U.S. SENATE PUBLIC WORKS COMMITTEE

CR

The packaging for a microwavable microwave dinner is programmed for a shelf life of maybe six months, a cook time of two minutes and a landfill dead-time of centuries.

DAVID WANN

CR

Increasingly, the world around us looks as if we hated it.

ALAN WATTS

CR

Filthy water cannot be washed.

WEST AFRICAN proverb

CR

As long as there are poor and minority areas to dump on, corporate America won't be serious about finding alternatives to the way toxic materials are produced and managed.

LEON WHITE

CR

The one process now going on that will take millions of years to correct is the loss of genetic and species diversity by the destruction of natural habitats. This is the folly our descendants are least likely to forgive us.

A very Faustian choice is upon us: whether to accept our corrosive and risky behavior as the unavoidable price of population and economic growth, or to take stock of ourselves and search for a new environmental ethic.

Pollution will do for the study of ecology what cancer did for molecular biology.

EDWARD O. WILSON

The Wilderness

One final paragraph of advice: Do not burn yourself out. Be as I am—a reluctant enthusiast . . . a part-time crusader, a half-hearted fanatic. Save the other half of yourselves and your lives for pleasure and adventure. It is not enough to fight for the land; it is even more important to enjoy it. While you can. While it is still there. So get out there and mess around with your friends, ramble out yonder and explore the forests, encounter the grizz, climb the mountains. Run the rivers, breathe deep of that yet sweet and lucid air, sit quietly for a while and contemplate the precious stillness, that lovely, mysterious and awesome space. Enjoy yourselves, keep your brain in your head and your head firmly attached to your body, the body active and alive, and I promise you this much: I promise you this one sweet victory over our enemies, over those deskbound people with their hearts in a safe deposit box and their eyes hypnotized by desk calculators. I promise you this: you will outlive the bastards.

Wilderness is not a luxury but a necessity of the human spirit.

The idea of wilderness needs no defense, it only needs defenders.

EDWARD ABBEY

CR

Wilderness is an anchor to windward. Knowing it is there, we can also know that we are still a rich nation, tending our resources as we should—not a people in despair searching every last nook and cranny of our land for a board of lumber, a barrel of oil, a blade of grass, or a tank of water.

SENATOR CLINTON P. ANDERSON

CR

You will find something more in woods than in books. Trees and stones will teach you that which you can never learn from masters.

ST. BERNARD

CR

Great things are done when men and mountains meet. This is not done by jostling in the street.

WILLIAM BLAKE

CR

Without wilderness, we will eventually lose the capacity to understand America. Our drive, our ruggedness, our unquenchable optimism and zeal and élan go back to the challenges of the untrammeled wilderness. Britain won its wars on the playing fields of Eton. America developed its mettle at the muddy gaps of the Cumberlands, in the swift rapids of its rivers, on the limitless reaches of its western plains, in the silent vastness of primeval forests, and in the blizzard-ridden passes of the Rockies and Coast ranges. If we lose wilderness, we lose forever the knowledge of what the world was and what it might, with understanding and loving husbandry, yet become. These are islands in time—with nothing to date them on the calendar of mankind. In these areas it is as though a person were looking backward into the ages and forward untold years. Here are bits of eternity, which have a preciousness beyond all accounting.

HARVEY BROOME

❧

You must have the bird in your heart before you can find it in the bush.

JOHN BURROUGHS

❧

Some national parks have long waiting lists for camping reservations. When you have to wait a year to sleep next to a tree, something is wrong.

GEORGE CARLIN

❧

It is good to realize that if love and peace can prevail on earth, and if we can teach our children to honor nature's gifts, the joys and beauties of the outdoors will be here forever.

JIMMY CARTER

❧

Nothing is more beautiful than the loveliness of the woods before sunrise.

GEORGE WASHINGTON CARVER

❧

I realized that Eastern thought had somewhat more compassion for all living things. Man was a form of life that in another reincarnation might possibly be a horsefly or a bird of paradise or a deer. So a man of such a faith, looking at animals, might be looking at old friends or ancestors. In the

East the wilderness has no evil connotation; it is thought of as an expression of the unity and harmony of the universe.

WILLIAM O. DOUGLAS

&

The action and tone of his statement leads me to conclude that Secretary Watt's idea of wilderness is a parking lot without lines.

DON EDWARDS

&

We have too long treated the natural world as an adversary rather than as a life-sustaining gift from the Almighty. If man has the genius to build, which he has, he must also have the ability and the responsibility to preserve.

GERALD R. FORD

&

The woods were made for the hunters of dreams,
The brooks for the fishers of song;
To the hunters who hunt for the gunless game
The streams and the woods belong.

SAM WALTER FOSS

&

In some mysterious way woods have never seemed to me to be static things. In physical terms, I move through them; yet in metaphysical ones, they seem to move through me.

JOHN FOWLES

&

Zoos are becoming facsimiles—or perhaps caricatures—of how animals once were in their natural habitat. If the right policies toward nature were pursued, we would need no zoos at all.

MICHAEL FOX

௧

The best remedy for those who are afraid, lonely or unhappy is to go outside, somewhere where they can be quiet, alone with the heavens, nature and God. Because only then does one feel that all is as it should be and that God wishes to see people happy, amidst the simple beauty of nature.

ANNE FRANK

௧

I remember a hundred lovely lakes, and recall the fragrant breath of pine and fir and cedar and poplar trees. The trail has strung upon it, as upon a thread of silk, opalescent dawns and saffron sunsets. It has given me blessed release from care and worry and the troubled thinking of our modern day. It has been a return to the primitive and the peaceful. Whenever the pressure of our complex city life thins my blood and benumbs my brain, I seek relief in the trail; and when I hear the coyote wailing to the yellow dawn, my cares fall from me—I am happy.

HAMLIN GARLAND

௧

Today the network of relationships linking the human race to itself and to the rest of the biosphere is so complex that all aspects affect all others to an extraordinary degree. Someone should be studying the whole system, however crudely that has to be done, because no gluing together of partial studies of a complex nonlinear system can give a good idea of the behavior of the whole.

MURRAY GELL-MANN

CR

The exquisite sight, sound, and smell of wilderness is many times more powerful if it is earned through physical achievement, if it comes at the end of a long and fatiguing trip for which vigorous good health is necessary. Practically speaking, this means that no one should be able to enter a wilderness by mechanical means.

GARRETT HARDIN

CR

Fieldes have eies and woods have eares.

JOHN HEYWOOD

CR

The spiritual uplift, the goodwill, cheerfulness and optimism that accompanies every expedition to the outdoors is the peculiar spirit that our people need in times of suspicion and doubt . . . No other organized joy has values comparable to the outdoor experience.

No prosaic description can portray the grandeur of 40 miles of rugged mountains rising beyond a placid lake in which each shadowy precipice and each purple gorge is reflected with a vividness that rivals the original.

HERBERT HOOVER

CR

What would the world be, once bereft
Of wet and wildness? Let them be left,
O let them be left, wildness and wet,
Long live the weeds and the wildness yet.

GERARD MANLEY HOPKINS

CR

To me a lush carpet of pine needles or spongy grass is more welcome than the most luxurious Persian rug.

HELEN KELLER

CR

The wilderness and the idea of wilderness is one of the permanent homes of the human spirit.

JOSEPH WOOD KRUTCH

CR

Wilderness is the raw material out of which man has hammered the artifact called civilization. Wilderness was never a homogenous raw material. It was very diverse. The differences in the product are known as cultures. The rich diversity of the world's cultures reflects a corresponding diversity. In the wilds that gave them birth.

ALDO LEOPOLD

CR

In wilderness I sense the miracle of life, and behind it our scientific accomplishments fade to trivia.

CHARLES A. LINDBERGH

CR

We have always had reluctance to see a tract of land which is empty of men as anything but a void. The "waste howling wilderness" of Deuteronomy is typical. The Oxford Dictionary defines wilderness as wild or uncultivated land which is occupied only by wild animals. Places not used by us are wastes. Areas not occupied by us are desolate. Could the desolation be in the soul of man?

JOHN A. LIVINGSTON

CR

Good heavens, of what uncostly material is our earthly happiness composed . . . if we only knew it. What incomes have we not had from a flower, and how unfailing are the dividends of the seasons.

JAMES RUSSELL LOWELL

ભ

Earth and sky, woods and fields, lakes and rivers, the mountain and the sea, are excellent schoolmasters, and teach some of us more than we can ever learn from books.

SIR JOHN LUBBOCK

ભ

God writes the gospel not in the Bible alone, but on trees and flowers and clouds and stars.

MARTIN LUTHER

ભ

There is nothing in which the birds differ more from man than the way in which they can build and yet leave a landscape as it was before.

ROBERT LYND

ભ

It should not be believed that all beings exist for the sake of the existence of man. On the contrary, all the other beings too have been intended for their own sakes and not for the sake of something else.

MAIMONIDES

ભ

There is just one hope of repulsing the tyrannical ambition of civilization to conquer every niche on the whole earth. That hope is the organization of spirited people who will fight for the freedom of the wilderness. In a civilization which requires most lives to be passed amid inordinate dissonance, pressure and intrusion, the chance of retiring now and then to the quietude and privacy of sylvan haunts becomes for some people a psychic necessity. The preservation of a few samples of undeveloped territory is one of the most clamant issues before us today. Just a few more years of hesitation and the only trace of that wilderness which has exerted such a fundamental influence in molding American character will lie in the musty pages of pioneer books. . . . To avoid this catastrophe demands immediate action.

ROBERT MARSHALL

જી

I only went out for a walk and finally concluded to stay out till sundown, for going out, I found, was really going in.

Climb the mountains and get their good tidings.
Nature's peace will flow into you as sunshine flows into trees.
The winds will blow their own freshness into you . . .
while cares will drop off like autumn leaves.

Let children walk with Nature, let them see the beautiful blendings and communions of death and life, their joyous inseparable unity, as taught in woods and meadows, plains and mountains and streams of our blessed star, and they will learn that death is stingless indeed, and as beautiful as life.

How glorious a greeting the sun gives the mountains!

JOHN MUIR

જી

Wilderness itself is the basis of all our civilization. I wonder if we have enough reverence for life to concede to wilderness the right to live on?

I hope that the United States of America is not so rich that she can afford to let these wildernesses pass by. Or so poor that she cannot afford to keep them.

MARGARET MURIE

ભ્ર

There is growing awareness of the beauty of country . . . a sincere desire to keep some of it for all time. People are beginning to value highly the fact that a river runs unimpeded for a distance. . . . They are beginning to obtain deep satisfaction from the fact that a herd of elk may be observed in back country, on ancestral ranges, where the Indians once hunted them. They are beginning to seek the healing relaxation that is possible in wild country. In short, they want it.

OLAUS J. MURIE

ભ્ર

Without knowing it, we utilize hundreds of products each day that owe their origin to wild animals and plants. Indeed our welfare is intimately tied up with the welfare of wildlife. Well may conservationists proclaim that by saving the lives of wild species, we may be saving our own.

NORMAN MYERS

ભ્ર

The wilderness holds answers to questions man has not yet learned to ask.

NANCY NEWHALL

ભ્ર

Wilderness to the people of America is a spiritual necessity, an antidote to the high pressure of modern life, a means of regaining serenity and equilibrium.

SIGURD F. OLSON

CR

Look at the trees, look at the birds, look at the clouds, look at the stars . . . and if you have eyes you will be able to see that the whole existence is joyful. Everything is simply happy. Trees are happy for no reason; they are not going to become prime ministers or presidents and they are not going to become rich and they will never have any bank balance. Look at the flowers—for no reason. It is simply unbelievable how happy flowers are.

OSHO

CR

The day I see a leaf is a marvel of a day.

KENNETH PATTON

CR

Climb up on some hill at sunrise. Everybody needs perspective once in a while, and you'll find it there.

ROBB SAGENDORPH

CR

And how should a beautiful, ignorant stream of water know it heads for an early release—out across the desert, running toward the Gulf, below sea level, to murmur its lullaby, and see the Imperial Valley rise out of burning sand with cotton blossoms, wheat, watermelons, roses, how should it know?

CARL SANDBURG

CR

Everything in nature is lyrical in its ideal essence, tragic in its fate, and comic in its existence.

GEORGE SANTAYANA

CR

Shall we, exploiting all our resources, reduce also every last bit of our wilderness to roadsides of easy convenience, and ourselves soften into an easy-going people deteriorating in luxury and ripening for the hardy conquerors of another century?

In the wilderness, we can get our bearings. We can keep from getting blinded in our great human success to the fact that we are part of the life of this planet and we would do well to keep our perspectives and keep in touch with some of the basic facts of life.

JOHN SAYLOR

CR

The more civilized man becomes, the more he needs and craves a great background of forest wildness, to which he may return like a contrite prodigal from the husks of an artificial life.

ELLEN BURNS SHERMAN

CR

Global warming threatens our health, our economy, our natural resources, and our children's future. It is clear we must act.

ELIOT SPITZER

CR

We simply need that wild country available to us, even if we never do more than drive to its edge and look in. For it can be a means of reassuring ourselves of our sanity as creatures, a part of the geography of hope.

Something will have gone out of us as a people if we ever let the remaining wilderness be destroyed; if we permit the

last virgin forests to be turned into comic books and plastic
cigarette cases; if we drive the few remaining members of the
wild species into zoos or to extinction; if we pollute the last
clear air and dirty the last clean streams and push our paved
roads through the last of the silence, so that never again will
Americans be free in their own country from the noise, the
exhausts, the stinks of human and automotive waste.

How much wilderness do the wilderness-lovers want? Ask
those who would mine and dig and cut and dam in such
sanctuary spots as these. The answer is easy: Enough so that
there will be in the years ahead a little relief, a little quiet, a
little relaxation, for any of our increasing millions who need
and want it.

WALLACE STEGNER

ভ

Trees are the earth's endless effort to speak to the listening
heaven.

RABINDRANATH TAGORE

ভ

Those who wish to pet and baby wild animals love them.
But those who respect their natures and wish to let them live
normal lives, love them more.

EDWIN WAY TEALE

ভ

Once you have heard the lark, known the swish of feet
through hill-top grass and smelt the earth made ready for the
seed, you are never again going to be fully happy about the
cities and towns that man carries like a crippling weight upon
his back.

GWYN THOMAS

ভ

In wildness is the preservation of the world.

HENRY DAVID THOREAU

ભ

Suburbia is where the developer bulldozes out the trees, then names the streets after them.

BILL VAUGHN

ભ

It is imperative to maintain portions of the wilderness untouched so that a tree will rot where it falls, a waterfall will pour its curve without generating electricity, a trumpeter swan may float on uncontaminated water—and moderns may at least see what their ancestors knew in their nerves and blood.

BERNAND DE VOTO

ભ

The magnificence of mountains, the serenity of nature— nothing is safe from the idiot marks of man's passing.

LOUDON WAINWRIGHT

ભ

Love is a powerful tool, and maybe, just maybe, before the last little town is corrupted and the last of the unroaded and undeveloped wildness is given over to dreams of profit, maybe it will be love, finally, love for the land for its own sake and for what it holds of beauty and joy and spiritual redemption that will make [wilderness] not a battlefield but a revelation.

T. H. WATKINS

ભ

Truly it may be said that the outside of a mountain is good for the inside of a man.

GEORGE WHERRY

❧

Without enough wilderness America will change. Democracy, with its myriad personalities and increasing sophistication, must be fibred and vitalized by regular contact with outdoor growths—animals, trees, sun warmth and free skies—or it will dwindle and pale.

WALT WHITMAN

❧

If you know wilderness in the way that you know love, you would be unwilling to let it go. . . . This is the story of our past and it will be the story of our future.

TERRY TEMPEST WILLIAMS